Why Socialism Sucks

An Engineer Analyzes Political Choice

ROBERT A. JONES

ISBN: 978-1-7376475-1-5 (Paperback)
ISBN: 978-1-7376475-0-8 (E-Book)

Library of Congress Control Number: 2021917717

Copy editing by David Brown

Cover and interior design by Andy Meaden meadencreative.com

First edition 2021

Contact the author at: rajpetn@gmail.com

For Henry, George, and John Mangialardi.

May the world you live in be better than the world I have lived in.

From the Cambridge Dictionary

suck verb (BE BAD)

[Intransitive] slang

We all thought the movie sucked.

CONTENTS

Preface 3

Introduction 7

ONE **Understanding Socialists** 9

Caring Progressives 10
The Socialist Mind 12
The Transgender Movement 16
Munchausen by Proxy Invades Venezuela 20
Munchausen by Proxy Invades Campus 21

TWO **Capitalism As It Is** 31

Capitalism Appeals to the Best Within Us 31
Capitalism and Freedom 35

THREE **Socialism As It Is** 39

Reality versus Fantasy 39
Socialism Means Repression 42
Socialism and Science 54
Socialism as Feudalism 61
An Intellectual Vacuum 72
Questioning Orthodoxy 82
White Privilege, Black Privilege, Yellow Privilege 85
Obsession with Race 89
Religion and Socialism 96

FOUR **Socialist Doomsaying and Global Warming** 105

How Marx Mimics Paul 106

Sea-level Hysteria 111

Politics, Not Science 118

The Power of Hatred 120

FIVE **Creating Art** 123

Art and Freedom 123

The Act of Creation 126

Poetry 131

Painting and Remembering 140

SIX **Creating Technology** 145

Engineering to the Rescue 147

The Case of the Deadly Dehydrator 150

Counting Annual Emissions 151

Let's Frack 158

SEVEN **Education of a Wandering Man** 161

Epilogue 179

Acknowledgements 181

Index 183

PREFACE

I have had the chance to observe socialism firsthand—the real socialism, as it is actually practiced, not the utopian fantasies of socialist dreamers.

I have only visited socialist lands; fortunately, I have never had to live under the grinding cruelty of a socialist regime. But these experiences motivated me to think about what I have seen. I now report my conclusions.

I draw on science, art, and my personal experience to contrast socialism and capitalism. Politics is the most human of affairs. Although dry scholarly treatments have their place, these matters are too vital to leave entirely to scholars and their sometimes abstruse works.

I've included a brief autobiography to give you a glimpse of my education and experience. My engineering degree and MBA, worldwide business travel, involvement in the arts, and extensive reading have all contributed to my worldview. With a background in technology, I am trained to pay attention to facts and to think about them logically. This book invites you to do the same.

Although you and I have different personal backgrounds, we can directly share some of the books and authors whose ideas have helped shape this book. They include:

The Road to Serfdom by F. A. Hayek. Everyone's first guide to understanding socialism.

"Surely You're Joking, Mr. Feynman!" Adventures of a Curious Character by Richard Feynman. A Nobel Laureate's first-person account of having fun while exercising the mind.

The Righteous Mind: Why Good People Are Divided by Politics and Religion by Jonathan Haidt. A fresh look at the psychology of political impulses.

Education of a Wandering Man: A Memoir by Louis L'Amour. Experience as the best teacher.

Guidelines for Hazard Evaluation Procedures by the Center for Chemical Process Safety and the American Institute of Chemical Engineers. A how-to book for keeping the world safe.

Fools, Frauds and Firebrands: Thinkers of the New Left by Roger Scruton. Leads us through the dank labyrinth of socialist thinking, damning socialists with their own words.

Drawing on the Right Side of the Brain by Betty Edwards. A book about learning to draw with a message for all.

Memoirs of the Second World War by Winston Churchill. As much about the man as the war.

An Introduction to Scientific Research by E. Bright Wilson, Jr. A justification for so-called deniers of unproven scientific claims.

Process Heat Transfer by Donald Q. Kern. It helped me a lot when I was starting my business.

Political Pilgrims: Western Intellectuals in Search of the Good Society by Paul Hollander. An exhaustive study of the socialist's intellectual stupidity—confirmation bias run amok.

The True Believer: Thoughts on the Nature of Mass Movements by Eric Hoffer. Should be required reading for its insight into those who would sacrifice themselves to a cause "to be rid of an unwanted self."

INTRODUCTION

Join me in the Berlin of the 1960s. Go behind the Wall to witness the desolation wrought by socialism. Ride with me on a coal-burning train from Beijing to Tsing Tao in the China of 1978, not very long after the end of Mao Zedong's Cultural Revolution and Great Leap Forward and his communist government's slaughter of 60 million Chinese. Sit in on my presentations to Chinese administrators and engineers and sympathize with their hope that the curse of socialism might soon disappear from China. Join me in Brezhnev-era Moscow. Enjoy the paranoid ambience of KGB Comrade Colonel Petrov's office at the Soviet Union Ministry of Foreign Affairs.

The path to wisdom is paved with reality.

The reality is that in the twentieth century, socialist ideas and practice caused tens of millions of deaths. Obviously, then, socialism has been discredited, right? Not so fast. In free, democratic nations all over the world, socialists are still busily trying to inflict socialism on the world. Not enough wealth and lives have yet been destroyed, apparently.

Why would anybody embrace the malignant dream that is socialism? In the first chapter, I consider the reasons for this irrational behavior.

Then, after explaining the nature of capitalism, I look at one of the key socialist messages, which amounts to "Trust me. We will get it right. Next time." Of course, being defiant of reason and reality, socialism never can and never will get it right. It is junk science from beginning to end.

The modern world was built on the basis of sound principles of engineering. Wisely employing some of these same principles can help us make better political decisions.

For example, before building something new and different, engineers often conduct a hazard analysis to examine potential risks to life and property. Such examination is the first step in reducing or eliminating such risks. Given how socialism has carelessly destroyed so many millions of lives, wouldn't it be eminently reasonable to consider the hazards of various political agendas before enacting them? Those likely to do nothing but destroy people and their achievements can be crossed off the list.

Socialism is dark and grim. But capitalism and human potential offer light and hope. After critiquing the socialists' global warming theology, I consider the nature of human creativity and how capitalism embodies that creativity and makes it possible.

ONE
UNDERSTANDING SOCIALISTS

In the late 1970s, my wife Gayle and I were living near Charleston, West Virginia when a minister at our church—where Gayle was organist and choir master—asked her to help him with an outreach ministry. Her job would be to play the piano at a Sunday evening service for mentally handicapped adults living at the Guthrie Center, a psychiatric facility on the outskirts of Charleston. They were being confined there because could not properly care for themselves.

Gayle loved providing music for the Guthrie inmates, always taking great pleasure in what was, for her, the end of a very long workday. They greatly enjoyed the hymn singing and fellowship as well and would always express their heartfelt appreciation.

Caring Progressives

One night, after Gayle had been visiting the facility for several years, she was told that it would be closing. It seemed that "caring" progressives had raised strong objections to confining people like the Guthrie inmates. Such confinement was cruel, they said; it trampled on their rights. With the success of progressive campaigns and court cases, Guthrie and places like it were deemed intolerable and were obliged to close their doors.

What was a victory for the progressives was not quite a victory for the alleged beneficiaries of their concern. Within days of the Guthrie shutdown, Gayle and I noticed a change in downtown Charleston. Former residents of the Guthrie facility who used to sing hymns with Gayle every Sunday night were now "homeless," living on the streets of Charleston. As the weeks went by, they looked worse and worse, dirtier and dirtier. Over time, they began to vanish—to be replaced by other "homeless" persons.

A tragedy? Not from the standpoint of true socialists. The sight of these unfortunate people supplies them with highly visible "evidence" that capitalism is a failure. The plight of the homeless is a perennial talking point for the left, who do nothing to ameliorate the problem when they have the chance, preferring to subsidize it instead. In San Francisco, for example, the city leaders are extremely tolerant of the way members of their homeless population spread feces on the town's streets and sidewalks. Such conduct has been widely reported in cities and appears to be most severe in this most Democratic of cities.

The residents of the Guthrie facility had been confined, but they had not been on the street. They had been clean and warm and safe. Although they had lacked freedom, they had not been one cold night away from freezing to death.

Socialist "caring" takes a grim toll on such souls. In 2018, there were 1,047 deaths among the 50,000 members of the Los Angeles homeless population. If the death rate of this population had been comparable to that of the United States population, there would have been only about 400 deaths among the LA homeless in that year. Being homeless may well have helped cause over 600 deaths, making the risk of death associated with homelessness in LA greater than that associated with infection by the COVID-19 virus. Data does not exist to show that similar death rates exist in other cities, but there is no reason to doubt that the problem is universal.

The situation is pretty good for progressives, though, for it has spawned an exciting new industry of advocacy for the homeless. Caring progressives raise money for, staff, and run this progressive-created industry. So not only can they congratulate themselves for helping these unfortunate homeless people, they can also enjoy the never-ending source of income that attends this demonstration of "caring." Their conduct brings to mind a condition called Munchausen syndrome by proxy (MSBP), defined by Michigan Medicine as "a mental health problem in which a caregiver makes up or causes an illness or injury in a person under his or her care, such as a child, an elderly adult, or a person who has a disability...[It] can involve any vulnerable person who has a caregiver." The condition has also been called Factitious Disorder Imposed on Another (FDIA). Whatever its name, MSBP has been recognized as a mental health problem for over 50 years.

What if socialism itself may be explained by MSBP or FDIA in the form of a mass psychogenic illness not unlike mass hysteria? The idea that the socialist political movement and its whole 150-year history is the result of a mental pathology sounds bizarre. Yet socialists do seem to imagine problems that afflict society that they then proceed to "solve" by inflicting outcomes even worse than those initially imagined problems.

Such a radical hypothesis must be subjected to detailed examination. It must be proved, or disproved, by careful study.

The Socialist Mind

In his excellent book *The Righteous Mind*, Jonathan Haidt proposes five moral foundations to explain how individuals choose their political philosophy. An individual's morality and political philosophy are grounded in these foundations.

What Haidt and others who have explored these ideas call Moral Foundations Theory is rooted in empirical observations. In this respect, unlike much of today's "social science," it emulates the hard sciences. Facts are observed. A theory is suggested to explain the observations, and this theory is tested against additional observations. Because it employs the scientific method, moral foundations theory is the most scientific and reasonable basis that I know of to explain a person's political philosophy. This theory is probably not yet in its final form. But the same may also be said of quantum mechanics or general relativity. Although perhaps not yet complete, they are useful in their current form.

What are the five moral foundations?

According to the care/harm foundation, the weak and children need our protection. This foundation underlies the virtues of kindness, gentleness, and nurturance. According to the fairness/cheating foundation, social cooperation is good and beneficial. This foundation underlies the ideas of justice, individual rights, and autonomy. According to the loyalty/betrayal foundation, we benefit from forming coalitions with others. This foundation underlies patriotism and self-sacrifice for the group. According to the authority/subversion foundation, we should defer to legitimate authority and respect traditions. This foundation promotes the virtues of leading and following. According to the sanctity/degradation foundation, the body is a temple that can be desecrated by immoral actions. This foundation is shaped by our disgust for contamination.

Haidt contends that members of the far left derive their world view almost entirely from two of these foundations: care/harm and fairness/cheating. Everyone knows how socialists protest the alleged unfairness of capitalism and insist that the downtrodden proletariat must be cared for. Haidt and his collaborators compiled a massive data set in order to relate the concern of individuals for each of the moral foundations to their personal politics. The graph below is a re-casting of Haidt's data. It shows that people who are "very liberal" (which certainly includes socialists) value caring and fairness much more than they value loyalty, authority, and sanctity. The graph opposite shows that conservatives have an evenly balanced appreciation of all of the moral foundations.

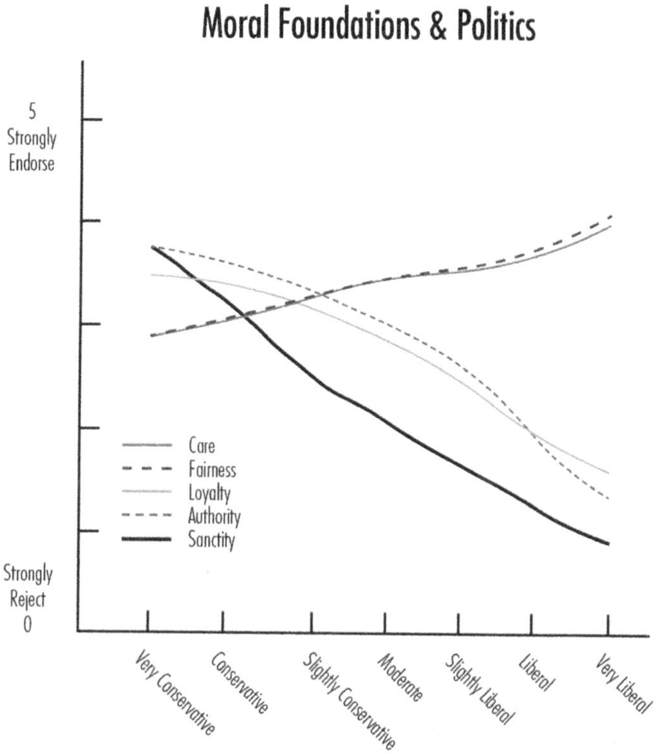

Socialists always stress how much they care for one oppressed group or another, and they demand equality above all else. Conservatives, by contrast, may give some weight to the care/harm and fairness/cheating foundations, but they are substantially concerned about the other moral foundations as well. They have a more balanced moral perspective.

Although Haidt acknowledges that socialists largely ignore three of his moral foundations, he does not acknowledge that they are, at least in this sense, unbalanced. Nor does he conclude that socialists are mentally unbalanced. We must. The consequences of a profoundly unbalanced mind are clear when we examine the actions of individuals concerned only with caring and fairness and who totally reject authority and sanctity. In their view, caring for the poor justifies stealing from the rich and giving the stolen property to the poor. After all, it is not "fair" that some are rich and some are poor, and these individuals defer to no authority that condemns theft.

If we wish to expand upon Haidt's understanding of the psychology of socialism, we must examine other examples of how socialists actually express their alleged caring and how they attempt to achieve their version of "fairness." Doing so will enable us to determine whether the behavior of socialists and progressives is indeed a form of Munchausen syndrome by proxy. The impassioned support that socialists give to all things transgender is a good place to begin this examination, since the harm that socialists do here—especially to children—is blatant.

The Transgender Movement

Individuals afflicted by the belief that they are actually members of the opposite sex deserve our sympathy and compassion. They cannot help who they are, and they are burdened by a mental illness that is beyond our current understanding. A further tragedy of their situation is that socialists are now exploiting their problems to further the socialist agenda.

This agenda even supports people who are actively harming helpless children by pushing them into identifying with the opposite sex. See, for example, the article "Trans Groups Under Fire for Huge Rise in Child Referrals" in the November 25, 2018 issue of *The Sunday Times* of London. "Referrals," here, involves telling kids between three and 11 years old that they can be any sex they choose, regardless of their actual genetically determined sex. In this way, adults are harming children by manipulating them into saying and doing things that are highly self-destructive. This latest explosion of child abuse recalls the false-memory syndrome of the 1980s, in which children were encouraged to report sexual abuse that had never happened to them. Cases in which adults imagine that children have an illness that they do not in fact have are examples of Munchausen syndrome by proxy.

Insofar as it resurrects the ideas of Trofim Lysenko, a hero of the Soviet Union, the transgender movement is a manifestation of neo-Soviet thought. Lysenkoism is defined by *Merriam-Webster* as "a biological doctrine asserting the fundamental influence of somatic and environmental factors on heredity in contradiction of orthodox genetics." Similarly, the transgender movement claims that human genes simply don't matter and that a person can "choose" his or her own gender—no matter which gender is in fact genetically expressed. Transgenderism is the new Lysenkoism.

Unfortunately, many doctors have embraced the transgender madness. Surgical "sex-change treatment" is a tragic replay of past failures. Decades ago, mental problems were often "cured" using a surgery called lobotomy, a butchery that consisted of cutting into the prefrontal lobe of the brain. Its most famous victim was Rosemary Kennedy, the sister of a future president, John Kennedy. Like all attempts to solve mental illness using surgery, this one, too, was a horrible failure. As *The Guardian* has reported ("The Forgotten Kennedy," August 12, 2009), the lobotomy "stopped Rosemary's violent behaviour but left her with the mental capacity of a two-year-old; incontinent and unable to speak intelligibly."

Imposing sex-change surgery on a child is rationalized by the claim that the child "wants" it. But the fact that a child wishes to harm himself or herself or has been cajoled into saying so is no justification for inflicting the harm. The same kind of manipulation has been used to rationalize female genital mutilation, since some young girls are convinced by their elders that submitting to the procedure is somehow a good thing. Islamic religion may provide a cover story, but forcing children to submit to such barbaric surgeries is an evil perpetrated by individuals who are themselves afflicted by Munchausen syndrome by proxy.

Admittedly, it is difficult to determine whether surgeons performing sex-change butchery are motivated by politics or by the revenue that they earn by conducting the surgery. However that may be, much of the motivation for the movement as a whole is undoubtedly political. Socialism.

According to the American Foundation for Suicide Prevention, "The prevalence of suicide attempts among respondents to the National Transgender Discrimination Survey (NTDS), conducted by the National Gay and Lesbian Task Force and National Center for Transgender Equality, is 41 percent, which vastly exceeds the 4.6 percent of the overall U.S. population who report a lifetime suicide attempt. . . ."

Laws are now being passed to ensure that transgender individuals are not discriminated against in employment. Since about 40 percent of all transgender individuals become suicidal at some point, being a passenger on an airplane with a transgender pilot is a risk comparable to that of flying on any airplane whose pilot is known to be suicidal. Many who knew the pilot of Malaysian Airline Flight 370, Zaharie Ahmad Shah, believed that he was clinically depressed when he killed himself, eleven other crew members, and 227 passengers. Instances of pilot suicide while the pilot is in the air do not always involve a terrorist motive.

Suicidal pilots are not the only risk. Transgender bus drivers and truck drivers also put us at risk.

Transgenderism is a mental illness. The failure by so-called "progressives" to recognize this is yet another example of how socialists rely on emotion rather than reason to analyze a situation. An objective analysis of the hazard involved suggests that transgender individuals should never be allowed to work in jobs where their suicidal tendencies may put lives at risk. Using transgender airline pilots to fly a plane makes only slightly better sense than using pilots who are legally blind.

Munchausen by Proxy Invades Venezuela

Claiming to care for the people of Venezuela, socialists destroyed its economy. They said that Venezuelan capitalism was *unfair* because some people in the country had more than others. To gain power, they lied about the depth of poverty in Venezuela, the richest country in South America at the time that the Chavez socialists first came to power in 1998. And after gaining power, their socialist "solutions" produced the same horrific poverty that they said had characterized Venezuela before they came along to rescue the country.

What happened in Venezuela is classic example of Munchausen syndrome by proxy. In socialist Venezuela, the socialist inmates are running the asylum. Remember that in this syndrome, in the words of the Cleveland Clinic, "a person acts as if an individual he or she is caring for has a physical or mental illness when the person is not really sick. The adult perpetrator... directly produces or lies about illness in another person under his or her care."

After ascending to power in the late 90s, Hugo Chavez placed Venezuela "in his care" in the name of the Bolivarian Revolution by promising to eliminate economic inequality. Socialists always demand equality for all, but this is a cry against nature: human beings are not in fact equal. Some are more athletic than others, more intelligent than others, more talented at painting or acting or music, etc. The outcomes of our lives necessarily differ because we begin life as differing individuals. We have different capabilities and talents and exercise different degrees of virtue in applying our gifts. Nor is luck evenly distributed.

To the extent that it can be approximated at all, real-world equality can be achieved only by the application of force. We will consider the issue of equality in greater depth later. But first we will examine how socialists are trying to change the nature of our institutions of higher learning.

Munchausen by Proxy Invades Campus

Socialists display Munchausen syndrome by proxy when they pretend that college students are dangerously fragile creatures who cannot endure the trauma caused by certain words—specifically, words that those students may disagree with or may feel offended by. Suddenly, safe spaces must be provided to protect college students from being exposed to disagreement, and trigger warnings must be provided prior to the uttering of any possibly sensitive material so that students may take steps to avoid hearing certain words or being exposed to certain ideas. These words and ideas are typically those that challenge socialist orthodoxy or that encourage debate on controversial social and intellectual issues. The same tender concern is not habitually extended to students offended by socialist twaddle.

Modern universities have never been perfect. However, once upon a time the imperfections of the university could be openly examined, and ways to improve it could be openly proposed and debated. Today, students are now deemed too fragile to endure such open debate.

Campus rape culture. University progressives have lately taken to devouring their own; even if you have otherwise impeccable progressive credentials—for example, are a devoted feminist—former fellow travelers will come down on you like a ton of bricks if you dare to question any part of the orthodoxy of the progressive Counter-Enlightenment.

Professor Laura Kipnis is a professor of media studies at Northwestern University with a history of support for women's rights.

When one of her colleagues, philosophy professor Peter Ludlow, was unfairly accused of sexual harassment, Kipnis wrote an essay in the *Chronicle of Higher Education* ("Sexual Paranoia Strikes Academe," February 27, 2015) that questioned the reasonableness of new campus regulations of relationships. She observed that women university graduate students were being treated like helpless children. "The melodramatic imagination's obsession with helpless victims and powerful predators is what's shaping the conversation of the moment, to the detriment of those whose interests are supposedly being protected, namely, students. The result? Students' sense of vulnerability is skyrocketing...[Our workplaces] are promulgating the crudest version of top-down power imaginable, recasting the professoriate as Snidely Whiplashes twirling our mustaches and students as helpless damsels tied to railroad tracks."

Kipnis argued that that treating adult women like children does a disservice to women in general; female graduate students are not universally weak and helpless and do not need overweening protection by university officials. In response to her essay, protesters demanded that she be fired. One of Ludlow's accusers teamed up with another graduate student to file Title IX complaint against her. According to the complaint, by stating her view of the situation the professor was guilty of "retaliatory behavior" and "creating a hostile environment."

The professor compounded her thought crime by writing a book about the progressive corruption of the university, *Unwanted Advances: Sexual Paranoia Comes to Campus*. The book includes an exhaustive description of the kangaroo-court Title IX process as it plays out on American campuses. One of the students discussed in the book brought a suit against Kipnis and the publisher for defamation and invasion of privacy. The same student joined with others to file yet another Title IX complaint against Kipnis.

As often happens in cases of Munchausen syndrome by proxy, the alleged cure of the alleged illness causes problems worse than the alleged illness itself. Such basic American ideals as a presumption of innocence are trashed and replaced by a presumption of guilt. As Kipnis puts in her book: "The problem is this: the more that 'safety' means lowering the bar for accusation-bringing, then the more of a magnet the process becomes, and has become, for anyone with an agenda, a grudge, a neurosis, and sometimes financial ambitions—payouts can be huge for a well-timed claim—and there's no adequate method for sorting legitimate from specious claims…It's not in administrators' interests to sort them: a campus's success in 'combatting sexual assault' is measured in increased accusations, which are closely tracked. By the way, complaints can increasingly be made anonymously—which is to say that witch hunt conditions are now an institutionalized feature of campus life."

The same bar-lowering is evident in the Me Too

movement's demand that any accusation of rape made by any woman against any man always be believed, whether the accusation is supported by evidence or not and despite the well-documented fact that some women have made false allegations of rape.

Unmentionable things. Open study and discourse in the universities were celebrated until progressives began discovering an intellectual "sickness" that had to be "cured." As a result, modern Western universities today showcase socialist, Counter-Enlightenment reaction. "How people feel is the most critical issue in our times" is the argument. "You must never, ever mention anything that might make any 'victim' feel bad!" Even mentioning a reality that "hurts a victim's feelings" is forbidden. The list of persons victimized by overt discussion of disagreeable ideas is almost endless, which is just what you'd expect given the nature of Munchausen syndrome by proxy.

Progressives demand that certain realities never be discussed or studied at the university. For example, any researcher who suggests that differences exist among the races—as manifested, for example, by the very high percentages of black players in the NBA and NFL—are a suitable subject for research is branded a racist. Charles Murray, co-author with Richard Herrnstein of the book *The Bell Curve: Intelligence and Class Structure in American Life*, which includes discussion of differences among races, has been accused of being a "racist" and a "white supremacist." In 2017, a gang of socialists acted on such prejudices by attacking Murray and Professor Allison Stanger at a Middlebury College when Murray came there to speak. Stanger ended up in the hospital.

Campus socialists and progressives enjoy the support of hate groups claiming to represent "victims." One such hate group is the Southern Poverty Law Center (SPLC). In its page on Murray, the SPLC charges Murray with using "racist pseudoscience and misleading statistics to argue that social inequality is caused by the genetic inferiority of the black and Latino communities, women and the poor." One won't find a judicious discussion of the pluses and minuses of the book at the SPLC web site.

In 2017, progressives on the University of Pennsylvania law school faculty attacked Professors Amy Wax and Larry Alexander for authoring an op-ed in the *Philadelphia Inquirer* ("Paying the Price for Breakdown of the Country's Bourgeois Culture," August 9, 2017). Wax and Alexander dared to suggest that rearing children within marriage, hard work, self-discipline, and respect for authority are vital to achieving a mature, prosperous society. Progressives leaped to label this racist, homophobic, sexist, and transphobic hate speech. Open inquiry is at risk at the University of Pennsylvania Carey Law School.

The end of diversity? The American universities of today resemble the German universities of the late 1930s. Conservatives in humanities departments are like Jews in Nazi-controlled universities. There aren't any, or there are only very few. The difference is that Jews were excluded from German universities by law, whereas conservatives are excluded from American humanities departments by dint of selective hiring and promotion. Just as "Irish Need Not Apply" signs were a common sight in late nineteenth-century America, so universities of twenty-first century American might as well be flaunting "Conservatives Need Not Apply" signs. Once hired, inmates of the university are often expected to toe the progressive party line notwithstanding any impulses in the direction of intellectual honesty and independence.

In 2017, organizers of a Day of Absence at Evergreen College in Washington state—by then an annual event—said they wanted white people in particular to stay off campus on that day. (Supposedly, the idea was to combat racism.) Biology Professor Bret Weinstein publicly objected, saying that there is

> a huge difference between a group or coalition deciding to voluntarily absent themselves from a shared space in order to highlight their vital and underappreciated roles (the theme of the Douglas Turner Ward play Day of Absence, as well as the recent Women's Day walkout), and a group encouraging another group to go away. The first is a forceful call to consciousness, which is, of course, crippling to the logic of oppression. The second is a show of force, and an act of oppression in and of itself...On a college campus, one's right to speak—or to be—must never be based on skin color.

Even though his comments endorsed the silliness of voluntarily absenting oneself from a "shared space" as a way to affirm one's progressive consciousness, this genuflection before the altar of progressivism failed to deflect the hyenas on campus. In addition to demands that he be fired, Weinstein was physically threatened. Evergreen did not fire him. But neither did it support his right to express himself freely. (Some of his students did strongly support him.) Weinstein sued the college for cultivating a "racially hostile and retaliatory work environment," and he and his wife eventually agreed to leave their posts in exchange for $500,000. A victory of sorts. Nevertheless, it has become increasingly clear that professors cannot with impunity risk open discourse and reasoned disagreement when progressive diktats are at issue.

Systematic exclusion of conservatives is officially sanctioned in elite American universities, who often don't even pretend to aspire to intellectual diversity among their professoriates.

"Why Are Nondiscrimination Statements Not Diverse?" asks David Rozado in a 2018 article for the Heterodox Academy site about a study by himself and Stephen Atkins. "Numerous studies have shown the severe lack of ideological diversity in universities across the U.S., with the vast majority of faculty leaning left of center. While most mission statements and diversity statements in the studied universities openly claim a commitment to protecting and fostering both demographic and viewpoint diversity, only a minority, 14 percent, explicitly protect viewpoint diversity in their non-discrimination statements. In contrast, archetypal categories of demographic diversity are overwhelmingly protected in non-discrimination statements. We find this disparity suspicious on the grounds of the vast underrepresentation of non-liberal viewpoints in the Academy...

"The results of our work suggest an easy step for educational institutions to wholeheartedly signal their welcoming of intellectual diversity. Namely, proactive institutional protection of viewpoint diversity in their nondiscrimination statements by extending to this type of diversity the same degree of protection already provided to archetypical categories of demographic diversity. Such a policy would communicate to internal and external stakeholders a strong institutional commitment to prevent discrimination and harassment based on ideas."

The problem with this prescription is that taking the conceptually "easy step" requires willingness to take the step. If university administrations don't see a problem—especially if they regard discrimination against conservatives as proper—they are unlikely to want to fix it.

Denial of discrimination is deeply rooted in academia. Harvard University is today defending itself from a lawsuit filed by Asian students who are trying to put a stop to Harvard's blatant discrimination against them—a discrimination motivated by progressive ideology.

TWO
CAPITALISM AS IT IS

Capitalism is rooted in every individual's dream of accomplishing something. These aspirations and the successes that flow from them enable us to survive and to improve our lives. As long as we can freely act in pursuit of our hopes and dreams, we will thrive.

Capitalism Appeals to the Best Within Us

Arguments in favor of capitalism—which may be broadly defined as the social system characterized by voluntary production and exchange and by protection of private property rights— typically focus on how capitalism benefits individuals and society. Arguments for socialism typically devolve into expressions of hatred for capitalism.

Even early societies that lacked any idea of capitalism or a free market had capitalists—in the general sense of any productive participants in markets and in the more specific sense of persons who invested capital (accumulated savings) in productive processes. These producers exercised their own initiative, made their own decisions, and depended on their own efforts to prosper. Some were farmers and some were craftsmen who created spears and bows and arrows for use in hunting. As societies grew and became more complex, technical skills such as weaving, sewing, sculpting, painting, and blacksmithing became common and provided these early capitalists with wealth, livelihoods, and the satisfaction of a job well done.

Capitalist enterprises were well-developed by the late Middle Ages. England's Domesday Book, commissioned in1085 by William the Conqueror, included six thousand water-driven and animal-driven mills among the taxable properties listed. The entrepreneurs who ground the nation's flour were forerunners of the Industrial Revolution. By the 1600s, the forerunners of modern corporations— joint-stock companies granted monopolies by the crown—were also being established.

In 1776, Adam Smith described in *The Wealth of Nations* how capitalism had improved the lives of *all* members of society. Smith described how an individual "generally... neither intends to promote the publick interest, nor knows how much he is promoting it...[B]y directing that industry in such a manner as its produce may be of the greatest value, he intends only his own gain, and he is in this, as in many other cases, led by an invisible hand to promote an end which was no part of his intention...By pursuing his own interest he frequently promotes that of the society more effectually than when he really intends to promote it." That is to say, the free actions of individuals, acting in their own interest, will also promote the interest of society generally. The total of best possible individual outcomes is the best possible social outcome.

Of course, Enlightenment thinkers like Smith did not invent capitalism. They merely observed it and thought about what they saw. They were among the first to understand capitalism and its greatness and to explain how capitalism works and delivers the greatest good for everyone.

The fundamental ethical requirement of free-market capitalism is simple. Grant true freedom to the individual, freedom that allows each individual to pursue his interests with only minimal intrusion by the state.

People must cooperate if society is to function. This cooperation requires that the individual enjoy the right of freedom of thought and action—so long as the freedom of one individual does not interfere with the equal right to freedom of other individuals. Under capitalism, individuals are free to act as they choose so long as their actions do not violate the rights of others.

The United States is based on the idea that individuals achieve their interest when they are free to act in accordance with their own judgment. This idea of trusting the people is the reason that the United States exists. It is why the Declaration of Independence speaks of "certain unalienable Rights," among which are "Life, Liberty and the pursuit of Happiness," rights that it is the purpose of government to secure; and why the U.S. Constitution speaks of establishing "this Constitution for the United States of America" in order to "secure the Blessings of Liberty to ourselves and our Posterity. . . ." These founding documents are about the rights of individuals to act in their own best interest.

The United States Constitution was established to ensure the maximum possible freedom of the individual and to severely limit the power of government. Men and women struggled and sometimes died to ensure that, as Abraham Lincoln put it, "government of the people, by the people, for the people, shall not perish from the earth." The practical implementation of our right to freedom includes the right to private property, the right to speak freely and openly about any subject, the right to assemble peacefully, the right to freely practice religion, the right to keep and bear arms, the right to due process, and the right to be free from unreasonable government action.

The United States was founded, in part, on the principle that no government has the right to run the lives of sane, adult, noncriminal human beings. (Of course, some citizens are not sane, adult, and noncriminal; and these persons require appropriately different treatment.) At its founding, the country was not perfect, but it was good enough to substantially achieve its republican vision while also providing for constitutional changes as needed through the amendment process.

Advocates of capitalism believe that people are happier and live better lives when they are free to do what they themselves decide to do (and are free to refrain from doing what they decide not to do). Socialists believe that people are happier and live better lives when they are forced to surrender their judgment and do what they are told to do by an allegedly all-knowing and all-powerful government. In the nineteenth century, American proponents of black slavery made similar arguments. The slaves were "happy," they argued. Lacking freedom, the slaves were protected from the consequences of any poor choices that they might have made had they been free. Apologists for slavery also argued that slaves were too stupid to make good choices for themselves. Today, socialists make similar arguments.

The American system of government is not perfect. It is only the most successful system of government ever devised.

Capitalism and Freedom

In a capitalist system, individuals are free to pursue their own economic ends.

The creators of the U.S. Constitution believed that in all important matters, the right of the individual to act as he chooses must take precedence over the wishes of the state. Of course, in some areas, commonly accepted rules that are a requirement of social cooperation preclude certain kinds of choices. Whether to drive on the left side or right side of a public road is not a decision we make every time our car leaves the driveway. The adoption of such conventions is consistent with the protection of our right to fundamental liberties; accepting them is tantamount to accepting the terms of a contract. (If roads were privatized, we can imagine that such rules, whether implicit or explicit, would be widely accepted on a contractual basis. Agreement to a contract means that one agrees to abide by its terms, a commitment that implies no loss of liberty.)

The world owes England everlasting gratitude for the concept of a bill of rights. The first document often mentioned in this context is the English Magna Carta of 1215. Some argue that the Magna Carta is not a true bill of rights inasmuch as it protected only the English nobility against royal abuses of power. The next document often mentioned is the Petition of Right passed by the English Parliament in 1628, which condemned unlawful imprisonment and provided that only Parliament had the right to approve taxes. The first document known as a bill of rights was passed by the English Parliament in 1689 under the formal title An Act Declaring the Rights and Liberties of the Subject and Settling the Succession of the Crown. A century later, some of its provisions were echoed in the U.S. Constitution.

The English Bill of Rights was passed at the beginning of the Enlightenment, or Age of Reason, an intellectual movement characterized by respect for the power of reason to achieve human happiness. This was the period during which the freedoms enjoyed by the common English people enabled England's economy to grow far beyond what the world had ever before seen, a growth that depended not on slavery or conquest but on hard work and innovation, the fruits of which were maximized by economic freedom.

You cannot have capitalism without freedom—including the freedom to succeed or fail.

THREE
SOCIALISM AS IT IS

Socialists have many complaints against the free market. They claim that capitalism is *unfair* in that it allegedly subverts democracy and individual freedoms and leads to exploitation of workers, imperialism and political oppression, and economic inequality. They claim that capitalism is *imperfect* in that it allegedly fosters inefficiency, irrationality, and unpredictability; leads to market failures and market instability; and permits unsustainable exploitation of natural resources.

When socialists say that capitalism is not perfect, they really mean that capitalism does not achieve the ideal (and idea) of perfection *promised* by socialist theory. Capitalism doesn't match their fantasy.

Reality versus Fantasy

Voltaire's aphorism that "the best is the enemy of the good" (often rendered "the perfect is the enemy of the good") is applicable here. The unrealistic and impossible are also the enemy of the good. Socialist theory sets an arbitrary standard of perfection and rejects anything that falls short (except for actual implementations of socialism). Socialists reject "good enough" capitalism, because it is not perfect by their definition—even if much-less-perfect socialist societies end up viciously subjugating and impoverishing the members of those societies, as we've seen in cases like Stalin's Soviet Union, Hitler's Third Reich, Castro's Cuba, and Chavez's Venezuela. The socialist ideal cannot be achieved—the promised paradise never arrives—and imposing a socialist regime can be attempted only by means of massive and destructive physical force that systematically throttles thought, initiative, and the creation of wealth.

The complaint that capitalism is "unfair" is infinitely elastic; it can mean anything that socialists want it to mean. A concern for "fairness" is not synonymous with a concern for justice, for justice pertains to how people morally evaluate and treat others. If one assesses others objectively and responds accordingly, rewarding good and punishing evil, one is being just. One cannot say that variations in nature or luck are "just" or "unjust," or that good weather here is "just" and bad weather there is "unjust"; but one can certainly complain about how unfortunate it is to suffer from bad genetic inheritance, bad luck, or bad weather. All of us are born with different capabilities and opportunities—which is surely "unfair" according to an abstract egalitarian standard that has no relationship to the world around us.

Socialist complaints about unfairness may be sincere, but they are also a tactic. Nazis used racism as a tactic to seize control; today's socialists respond to the alleged imperfections of capitalism with demands for "equality" and "fairness" and "common sense," with the goal of replacing capitalism with socialism. To justify his Bolivarian Revolution—a revolution now enforced by starvation and murder—Venezuelan socialist Hugo Chavez proclaimed that he would "reduce social disparities." In the Ukraine, Stalin starved millions to death.

When they lack absolute power, socialists destroy individual freedom more gradually, but as quickly as they can. A new law passes, restricting some individual right "for the good of the people." More restrictions are enacted. Then more, and still more. The fact that the worst consequences may arrive the day after tomorrow is no reason for lovers of freedom to refrain from acting today to prevent those eventual consequences. Remembering the words of Martin Niemöller about the Nazi takeover of Germany, freedom-loving people should oppose all restrictions on basic human freedoms: "First they came for the socialists, and I did not speak out—because I was not a socialist. Then they came for the trade unionists, and I did not speak out—because I was not a trade unionist. Then they came for the Jews, and I did not speak out—because I was not a Jew. Then they came for me—and there was no one left to speak for me." (There is an irony here that Niemöller himself apparently did not grasp; for the Nazis, although they railed against "Bolshevik socialism," were themselves socialists of a slightly different ilk—"national" socialists—and made no secret of it.)

Socialism Means Repression

The socialism described by socialists is a fantasy.

Perhaps the most egregious example of the fantasizing—lying—was perpetrated by Walter Duranty, a "reporter" for *The New York Times* who received a Pulitzer Prize for his accounts of the Union of Soviet Socialist Republics in the wake of the Bolshevik victory in the Russian Civil War. The Moscow Bureau Chief for the *Times* from 1922 to 1936, Duranty claimed that people sent to the death camps of the Soviet Gulag went there by their own choice. He also denied the existence of the Ukraine famine of 1932 to 1933, when three million to seven million Ukrainians were intentionally starved to death by the Russian regime. According to one of his articles ("Russians Hungry, But Not Starving," March 31, 1933), "Conditions are definitely bad in certain sections—the Ukraine, North Caucasus and Lower Volga. The rest of the country is on short rations but nothing worse. These conditions are bad, but there is no famine."

Full socialism exists in the real world only under brutally repressive regimes run for the benefit of a socialist Führer and his supporting elite. The only "socialist successes" that one hears about today are not really socialist. They may be the capitalist Scandinavian nations, which tax the earnings of workers in capitalist enterprises in order to support an extensive social welfare system. These earnings flow not from state-owned and state-directed enterprises, but from privately owned businesses.

What is socialism? *Merriam-Webster* defines it as "a system of society or group living in which there is no private property," a system which could presumably be set up voluntarily and with only a few people. Another definition—the one relevant to our discussion—is "a system or condition of society in which the means of production are owned and controlled by the state." The relevant definition of communism is "a totalitarian system of government in which a single authoritarian party controls state-owned means of production." (These "state-owned means of production" presumably constitute all or almost all the means of production in the society, excluding any black-market production conducted by evading state control.) Whether we're talking about socialism generally or the totalitarian one-party version, communism, the point is that under socialism the state, not private owners, control the means of production. There is no way to do this except by the sweeping use of physical force, including mass expropriation of private property as a socialist state comes into existence.

In a fully socialist society, the state has supreme authority and individuals are reduced to servants: the individual exists to serve the state. He must submit to the will of that state's führer, party chairman, or president-for-life.

The theory of socialism is a hodgepodge of often conflicting assertions and assumptions, but a key feature is indifference to reality. By downplaying or ignoring the painful realities of Soviet Russia, Nazi Germany, communist China, communist Cuba, or communist Venezuela, socialist theorists neglect the wisdom of physicist Richard Feynman: "It doesn't matter how beautiful your theory is; it doesn't matter how smart you are. If your theory doesn't agree with experiment, it's wrong." He also said: "Reality cannot be fooled." The real-world socialist experiments of countries like Soviet Russia and communist China proved by their abject failures that socialist theory is nonsense. Reality cannot be fooled by socialist theory.

The implementation of capitalism can be complicated. It can be difficult to balance the need for some governmental rules against the need to give individuals all possible freedom. The implementation of socialism is simple. The socialist führer issues the orders and the citizens either follow the orders or are imprisoned or killed. China today has "re-education camps" holding millions of Uighur Muslims. Capitalism needs no such oppressive institutions. Freedom does need not be *imposed* on a population. In a free society, government force is wielded only against those who coercively violate the rights of others, not against those who peacefully exercise their own rights.

China has undergone spectacular changes since Deng Xiaoping seized control of the government in 1978. Under Mao, the Chinese often starved; now many are well-fed. Under Mao, only ranking party officials had private telephones (or private toilets); now many Chinese have access to the Internet (albeit a party-controlled Internet). Having watched socialism fail in the country, Deng understood that socialism doesn't work, so he (partly) dismantled it. The government tried to bring capitalism to China, but it has done so very inconsistently. This quasi-capitalism has permitted more economic production than was possible under full socialism. China transformed itself and achieved vast improvements, but the Chinese people are not free to fully exploit their creativity and industriousness. Moreover, a capitalist society requires freedom that extends beyond economic relationships. The Chinese government has been giving lip service to individual freedoms while denying citizens the freedoms necessary for capitalism to flourish in the long run.

Adam Smith used the metaphor of the "invisible hand" to dramatize the fact that the free actions of individuals acting in their own best interests also produce benefits for society more generally. Except in very limited circumstances, the Chinese government does not allow individual freedom of action; yet it wants the benefits of capitalism. This makes China a schizophrenic society, subject to a nagging, never-ending instability. The very visible hand of the Chinese Communist Party tells all Chinese citizens: "Follow our orders or we'll imprison you."

State capitalism is a form of socialism based on recognition of the fact that socialism is a failure. But the system is still one that permits the existence of only a single political party, whose leader has more or less absolute power. This party routinely crushes any dissent.

The communism of Mao Zedong (who ruled from 1949 to 1976) was replaced by state capitalism because communism as actually practiced had destroyed much of the Chinese economy. Industry was a shambles, and Chinese agriculture was having great difficulty feeding China's population. China's state capitalism exists for the same reason that Hitler's regime existed: economics. The hardships caused by the Great Depression of the 1930s—a result of statist intervention in markets—enabled Hitler to overthrow capitalism and institute his own version of state capitalism. Because of the hardships caused by full-scale communism, China now has a social system much like that of Nazi Germany: a pseudo-capitalist and totalitarian one-party state.

Consider other similarities of China to Nazi Germany, like the massive infrastructure projects now being touted by China's leadership. In particular, the multi-trillion-dollar Belt and Road Action Initiative echoes Nazi Germany's autobahn construction of the 1930s. The justification for these massive expenditures, whether Chinese or German, comes down to job creation and nationalistic, pride-inducing boasts.

Another parallel is Germany's *lebensraum* and China's nine-dash line declaration, by which China claims much of the South China Sea. This claim was first made in 1947 by the Nationalist Chinese government. Under Mao, the communists took up the claim but excluded the Tonkin Gulf in deference to their brother communists of North Vietnam. Today, however, China is vigorously pressing the entire claim, with a massive island-building program and threats of military action against its neighbors. Germany's waging of WWII in part to gain "living space" for colonial expansion and China's current agenda are alarmingly similar. Indeed, both are declarations of a colonial intent and both were fully acted upon decades after being proposed by an earlier, hated regime. Imperial Germany first announced a policy of *lebensraum* in the late nineteenth century, and the Nationalist Chinese first proposed the nine-dash line in 1948.

Committed socialists have often contended that communism and Nazism are opposites. In fact, they are minor variants of each other, with different emphases but the same core totalitarian doctrines and practices. Unlike the Nazis, Chinese leaders have not openly declared the superiority of their own race. They may well be tempted to do so, since in some social contexts, claims of racial superiority can be visceral and unifying. But making such claims would be too obviously Nazi-like. And given the homicidal racism that the Japanese inflicted upon China during WWII—still well remembered—racist calls to action would disgust many Chinese.

In China, we see socialism in a relatively realistic form. China had destroyed much of its native capitalism by such programs as the Great Leap Forward (1959-1962), which wiped out small-scale farming for the sake of a misguided, coercively imposed communal system of farming that resulted in the starvation of an estimated 20 million people. As Mao lay dying, at least some Chinese leaders could see that the economic rationale for socialism was nonsense.

Modern Chinese society is a dictatorship that to some extent rejects central planning in favor of the material incentives of capitalism. The difference between Nazi Germany and contemporary China is that the Nazis jumped directly to the sort of political arrangement that China has now. Mao had destroyed markets and thereby nearly destroyed China itself. The new China maintains a veneer of capitalism while retaining pervasive police-state controls. Individual freedoms exist in today's China in the same way that individual freedoms existed in Nazi Germany: grudgingly at best. Hitler needed German enterprise, for he had wars to fight and nations to conquer. Like Nazi Germany, modern China has a somewhat functioning society with an elite ruling class in complete political control. Xi Jinping is the king of China in all but name.

With no guiding principles other than the pragmatic one that "what we are doing works right now," the sort of feudalism being practiced by the Chinese Communist Party does not even have a fig leaf like the divine right of kings to rationalize its shortcomings. For this reason, we may hope that the Chinese people will not allow themselves to be led further down a Nazi-like path. But the prospects are not encouraging. The country's government has no sound guiding principles. It rejects the fundamental principle of capitalism that every individual be protected in the exercise of his right to think and act as he chooses. Things are so much better now than they were under Mao that most of the Chinese people have been willing to go along with current regime. As fewer and fewer people remember the horrors of Mao, though, controlling the citizenry may become more difficult.

The situation in the United States is different. The U.S. has a constitution with a bill of rights. American political leaders can flourish the document and say: "This is who we are. This is what we believe. Love it. Fight for it, if necessary." This American call to action inspires opposition to enemies foreign and domestic. It gives the United States an ultimate source of strength that China lacks. The Constitution is also the means of defeating all domestic terrorists, whether they call themselves QAnon or Antifa. But the American call to action will fail if Americans become apathetic. As Ronald Reagan said, "Freedom is always one generation away from extinction."

Socialists blame all of society's ills or all challenges of life on capitalism, including problems inherent in human nature or in the requirements of survival as such. In response, socialists create their own quasi-secular religion: national socialism, communism, the Bolivarian Revolution, the Cuban Revolution. Whatever it is called, socialists simultaneously deny its fundamentally religious character (Marx called his socialism "scientific socialism") while worshiping its scriptures. Both fundamentalist Christians and dedicated socialists believe in a sort of verbal plenary inspiration for their holy writings. The former believe that God dictated the scriptures, while the latter trust in the writings of Marx and such lesser deities as Engels and Althusser.

The Counter-Enlightenment and its kindred evils begat modern socialism. Many people are familiar with the movement called Romanticism and the Romantic Era. The term Counter-Enlightenment may more accurately describe this movement, which appeared around the time of the French Revolution near the end of the eighteenth century. It may be no accident that the hopeless and bloody failure of the French Revolution came in its wake.

The thinkers of the Counter-Enlightenment rejected the philosophes' admiration of rational thought in favor of human emotion and intuition. Their movement was a reaction against modernity in general and the Industrial Revolution in particular. People moving into gloomy and cramped quarters in fast-growing cities helped to fuel a sudden appreciation for nature. Distaste for these cities, factories, and technology in general inspired early socialists to adopt the notion of the "noble savage" in order to glorify the purported perfections of primitive life and to reject the modern world.

Admiration of the noble savage is alive and well among today's progressives. These days, white progressives admire people of color who are rioting and looting and burning while flagellating themselves and denouncing "white privilege." We see this admiration of the noble savage in the idealization of the Black Lives Matter movement. But it's unlikely that today's progressives understand this connection or their own implicit racism.

The term "progressive" is a misleading name for this way of thinking; "reactionary" is more apt, since progressives are reacting against key principles of the Enlightenment. Progressives would have you believe that they are leading the way to a post-colonial, post-modernist world, when they are really leading us to a pre-Enlightenment world that more closely resembles feudalism.

Progressives expose their fascist and socialist roots when they attack anyone who questions climate-change orthodoxy. Their cries of "Denier!" in response to the specter of fundamental discussion and debate reveal their quasi-fascist anti-intellectualism, best described by Umberto Eco, who experienced fascism firsthand, in his essay "Ur-Fascism" ("Primitive Fascism"): "The critical spirit makes distinctions, and to distinguish is a sign of modernism. In modern culture the scientific community praises disagreement as a way to improve knowledge. For Ur-Fascism, disagreement is treason."

Karl Marx observed or fabricated the imperfections of capitalism and dreamed up a scheme to "solve" its problems. His followers contended that socialism is a "science" the perfection of which will eventually consign capitalism to the ash heap of history. But socialist theory was created without any reference to the real world, and it has failed in every society that has sought to implement it. Socialism has succeeded only as a means of seizing power and stomping on the rights of innocent people, with its failures rationalized or obscured by mounds of unintelligible theoretical complexity.

Socialists are utopians. If society is not perfect by whatever standard they choose to apply (however unreasonable), they guarantee that they will make it perfect. This is why today's socialists call themselves "progressive." But the term "neo-Soviet" is more apt. The Soviet socialists believed in the tenability of their theory until the reality of its implementation destroyed the Soviet Union. Stalin murdered tens of millions of Russians (and Mao murdered tens of millions of Chinese), but today's neo-Soviets simply ignore that inconvenient truth. They dispute that these murders were perpetrated to advance socialism dogma, and they will even explain that Stalin and Mao were not "real" socialists.

This misshapen worship of mankind motivates the attempts of today's socialists to remake society into a form that they would call perfect. Since they believe that society can be made perfect, signs of imperfection drive them crazy. Observe the socialists as they criticize the Founding Fathers of the United States. The unforgivable sin of the Founding Fathers, in their view, is their imperfection, their failure to entirely escape the limitations of their social and historical context. Criticism of Thomas Jefferson's ownership of slaves is one example. The facts that Jefferson was in many respects a wonderful and brilliant man—that the philosophy of individual rights enshrined in his Declaration of Independence led ultimately to the destruction of slavery—that slavery was a worldwide and ancient blight, hardly unique to the North America of the eighteenth and preceding centuries—are ignored. Such considerations are not carefully weighed in the minds and words of those who only anathematize the Founders.

The modern socialist or neo-Soviet movement is a hodgepodge of sometimes contradictory ideas. To understand it requires a multi-faceted examination.

Socialism and Science

Isaac Newton's revolutionary understanding of motion and gravity, widely accepted as the unchallengeable truth about how the universe works, ruled the eighteenth and nineteenth centuries. The clockwork universe he described answered many important questions, and its success undergirded the belief that everything was now, or soon would be, predictable and knowable. Anyone claiming otherwise was surely either a fool or troublemaker.

Seeing the certainty and perfection of Newtonian physics, Marx, Engels, and other socialists developed a "scientific" theory of economics and society that in their view represented the same kind of perfection as Newton's physics. The successors of Marx said, in effect: "Science is good. Governments must employ science to run the affairs of nations. Marx gives us the science that enables us to understand the future and to create plans to run the economy so as to achieve the best possible future for mankind."

The twentieth century was unkind both to Newtonian physics and to Marxist economics, but in very different ways. The success of quantum mechanics and the theories of special and general relativity showed that Newtonian physics is incomplete, because it did not take into account all possible conditions; nevertheless, pitchers, astronomers, and rocket scientists still rely on Newton's insights. But what of socialism's great "successes" in the twentieth century, like Stalin's Soviet Union, Hitler's Third Reich, the People's Republic of China, the Democratic People's Republic of Korea, Castro's Cuba, the Bolivarian Revolution of Venezuela? Each of these states employed socialist "science," and each failed spectacularly, providing further proof that Marxist economics is nonsense. The intricate plans that controlled these economies all failed to deliver the promised economic growth and became a mere excuse for maintaining murderous elites in power. If a scientific theory is true, it must yield in practice the results promised by that theory—a test that socialism fails.

As they dance in uncertain, unpredictable patterns, the individual units (i.e., people) of all economies are more like unpredictable subatomic particles than the predictable orbiting of the planets. The five-year plans of the central planners had no chance of successfully modeling the actions of myriad unpredictable human individuals. Nevertheless, many of today's socialists still religiously hew to their view of society as consisting of readily manipulatable automatons, despite the endless real-world failures of their planning.

In 1977, during a business trip to Moscow, seat of power of the Soviet communist empire, the reality of government planning slapped me in the face. With the goal of selling polyethylene manufacturing technology, Unipol, to the Soviets, my group met with members of Techmashimport, the Soviet department responsible for buying Western technology. After the meeting, a Soviet engineer named Yuri, probably a junior member of the KGB, offered to drive me back to the Intourist, the mid-Moscow hotel where I was staying.

Yuri suggested we have drinks and talk. Knowing that he was surely under KGB orders, I assumed that he had not been prompted by a simple desire to be friendly. In any case, when we reached our destination, he got out, removed the windshield wipers, and placed them inside the car before closing and locking the doors.

Not wishing to offend, I did not ask him about such odd behavior. Later, however, I learned from a British consultant traveling with our group that Soviet planners had forgotten to include windshield wipers in the latest Soviet plan. Because it was impossible to buy them anywhere, used windshield wiper blades were now very valuable. Leaving them outside your car guaranteed that they would be stolen—even in broad daylight and in the middle of Moscow.

So much for government planning.

That is not to say that Soviet five-year plans were poorly thought out. They may well have been the result of hard work by serious, bright, dedicated people. But they were nonetheless worthless, for they were based on the assumption that "rational," long-range plans to control society stand any chance of succeeding. One-size-fits-all central plans cannot predict, emulate, or substitute for the incentives and choices of voluntary human action, or for individual knowledge and judgment of local circumstances, no matter how cleverly conceived those plans may be.

The guaranteed failure of such central planning does not mean that human events or the economic life of a nation cannot be significantly *controlled*, however. Modern control system technology suggests how such control can be effected. During the Middle Ages, the control system was feudalism; control was exercised by kings through their vassals. Socialist societies have socialist theory and socialist practice. Capitalist societies have the free market, free individuals, and a legal regime upholding rights of property and contract.

Capitalism uses feedback control. Socialism uses feed-forward control.

No socialists in the 1800s understood control-system technology; if they had, they might have realized that their "science" was nonsense.

A microwave oven relies on *feed-forward control.* When you place a cold cup of coffee in the machine and press, for example, a button labeled "beverage," after a short time a beep lets you know that your coffee is now adequately heated. No measurements are taken to determine whether this is in fact the case; once the button has been pressed, the machine simply applies a predetermined amount of power for a predetermined period of time. Everything works perfectly as long as the planned-for conditions have been satisfied; e.g., the cup must have been filled with cold liquid coffee and not with iced coffee.

The drawback of feed-forward control systems is that they cannot cope with conditions different from those that were planned for when the machine was designed. In the case of a microwave oven, the programming permits only a specific amount of heat to be input during the chosen cycle, an amount sufficient to properly heat a specific item. Once you push a button, the rate of heat energy input and the length of time the item will be heated is set for you; that's it; whether your coffee begins as pure liquid or has ice floating in it, only the predetermined amount of heat will be applied unless you stop the process and choose new settings.

Suppose you install a feed-forward system to control the temperature of your home. You would first gather the temperature records for your location to determine the average temperatures for each hour of each day of the year, then determine how many minutes per hour your heating (or cooling) system must be switched on and off to maintain a desired internal temperature. If the outside temperature is much cooler than what you planned for, your home will be too cool; if much warmer than what you planned for, it will be too hot.

A centrally planned economy is a massive feed-forward control system. As in the case of the Soviet Union's five-year plans, everything is planned up front. Even if the planners' knowledge of current real-world conditions is completely accurate and exhaustive, the world begins to deviate from the conditions assumed in the plan just as soon as it begins to be implemented. The more time passes, the more the world deviates from the conditions necessary for the plan to succeed. Failure is inevitable even if the plans are perfectly constructed to fit the starting conditions. The actual situation is even worse, for normal human errors and lack of omniscience guarantee that the plan will contain flaws even on its own terms, as exemplified by the problem of the missing windshield wipers.

Socialist economies always fail in practice. The most recent major example is Venezuela. Venezuela has imploded quickly, since it started with a weak capitalist economy. It would take a socialist United States a bit longer to implode, given its stronger capitalist economy. But implode it would if subjected to the same strangling oversight.

Organisms are governed by *feedback control.* For example, in organisms of sufficient complexity, sensory organs receive inputs from the surrounding environment, and these inputs have effects on the nervous system and brain. The brain makes decisions that are appropriate to the situation, and the organism acts accordingly. A child feels pain and yanks his hand from a hot stove; a frog sees a fly and eats it. In every case, some outside stimuli are detected and some kind of action is taken in response.

Most of our homes have feedback temperature control systems that regulate the temperature inside the home automatically. These gadgets, thermostats, enable us to choose the temperature—the set point—that we want to maintain in the house. When the temperature has fallen a degree or two below the set point, the thermostat turns on the heating system, which heats the house and raises the temperature. When the thermostat detects that the temperature has risen a degree or two above the set-point temperature, the thermostat turns off the heating system. This regulatory mechanism is a simple form of feedback control.

Feedback control will never provide "perfect" temperature control; the temperature of your home fluctuates above and below the set point that you have selected as the optimal temperature. Similarly, the prices that reflect the outcomes of transactions in the free market never provide "perfect" economic outcomes. They don't prevent entrepreneurial error, for instance. But what prices and other aspects of the free market and free society provide is the means for economic actors to constantly adjust and improve their responses to real conditions.

In an unhampered market, when demand for a good or service increases, the price of that good or service tends to increase. If the price is too high with respect to real demand, demand will fall. Producers increase or decrease their production and prices in response to changes in demand and supply (the demand for and supply of their own goods as well as of complementary goods and factors of production). Both producers and consumers respond to prices established by market exchanges. This continuous market process is not perfect at regulating the supply of every good or service of the entire economy, but it is good enough and in fact cannot be improved upon.

Socialism as Feudalism

Socialism is a modern form of feudalism. The aristocrats of the feudal era regarded the institutions governing their society as fair and nearly perfect. It had to be so, in their view, for feudalism had been ordained by God. According to the principle of the divine right of kings, to occupy the top rung of feudal society was a God-given right. God himself had conferred upon the king the right of total control. The king's vassals, the feudal aristocrats who carried out his wishes, were each in turn also doing God's work, and they were rewarded by the king for doing so with fiefs that usually consisted of land but might also include perquisites like hunting or farming rights.

Socialism is feudalism in disguise. Instead of God, it is socialist theory that bestows socialists with control of society. Socialists worship the saints of socialist theory and practice, men like François Babeuf, Karl Marx, Eugene Debs, Vladimir Lenin, Mao Zedong,

Fidel Castro, Hugo Chavez. To enact his will, the führer, first secretary, or president for life of a socialist society appoints socialist aristocrats who receive jobs, property, and perks as their earthly reward for faithful service. Only members of the ruling socialist party—the bureaucrats and secret police operatives—can be members of this aristocracy.

Socialists assure us that the means of production really belong to "the people." Feudal kings and socialist führers alike claim that their purpose is to care for "the people." They say that their motives are selfless and that their highest devotion is to the public welfare. Why else would God or Marxist doctrine have given them such power over everybody?

In the Soviet Union, the socialist aristocracy was called the nomenklatura. Virtually all of them were members of the Communist Party. The nomenklatura was called a "new class" by Milovan Đilas, himself a longtime communist who became a harsh critic of communism. In communist countries, the socialist aristocracy is hereditary, with membership in the party being passed on to family members as a feudal right of socialist power. In China, the children of high Chinese Communist Party officials are secretly referred to as "princelings."

The saying that the more things change, the more they stay the same is certainly true of Russia, where heads of states are tantamount to the autocratic czars that the communists displaced. Russia's "democracy" is only a cover story; the Russia of today, like the Soviet Union of yesterday and the Czarist Russia of centuries ago, is a feudal society. Although Vladimir Putin's title happens to be "president" rather than "czar," it's a distinction without a difference. As Ronald Reagan observed, the Bolshevik revolution "simply replaced an inherited aristocracy with a self-appointed one, the Soviet leadership."

In North Korea, the justification for the ruling Kim dynasty is not so much socialist theory as the divine right of kings. The Kims are said to be descended from Tan'gun, the mythical founder of the Korean nation still worshipped today, and the holiday celebrating him is the only one observed in both Koreas. North Korea's shift from a socialist justification to a religious justification is another indication that communism is a form of feudalism.

Only war or revolution can end the family power of the Kims. Capitalism, on the other hand, is unkind to hereditary elites. Capitalist democracies have no mechanism other than wealth for bequeathing family power. But this wealth is not maintained automatically, typically being dissipated by the generations that inherit it. The proverbial summary of the situation is "shirtsleeves to shirtsleeves in three generations"; and when the money goes, the power goes too. This is why old-money families and power-elite families—the Vanderbilts, Rockefellers, and the Kennedys—over time tend to fade in prominence in capitalist countries. The Bushes and Clintons are seeing their power fade rapidly as well, since the only power they had is that which attends elective office.

Socialists complain about the "unfair" concentration of wealth in the hands of first-generation super-rich achievers like Bill Gates, Mark Zuckerberg, and Warren Buffet. But these people became rich by their own efforts, providing goods or services desired by the people, i.e., their customers. The customers created these mega-rich businessmen. When socialists rail against the so-called one percent, they imply that people are too dumb to know what they are doing as they engage in value-for-value exchanges in the market.

The communists and the Nazis both call themselves socialists. The word "Nazi" is short for "National Socialist German Workers Party," and the Russian communists created the Union of Soviet Socialist Republics. Socialists sometimes contend that the communists of Stalin's Soviet Union and Mao's China were not "real" socialists or that they "didn't get socialism right." The Nazis, too, were not "real" socialists in the view of some socialists. Yet Stalin, Mao, and Hitler all regarded themselves as socialists.

Like all socialists, Hitler and his Nazis hated capitalism. After Hitler assumed dictatorial control of the Nazi Party in 1921, everything the Nazi Party said and published reflected the thoughts of Adolph Hitler.

("Ein Volk, Ein Reich, Ein Führer" was a popular slogan of the party.) According to a 1926 Nazi pamphlet by Gregor Strasser, "We are Socialists, we are enemies, mortal enemies of the present capitalist economic system with its exploitation of the economically weak, with its injustice in wages, with its immoral evaluation of individuals according to wealth and money instead of responsibility and achievement, and we are determined under all circumstances to abolish this system!" Nazi minister of propaganda Joseph Goebbels declared that the bourgeoisie must "yield to the working class...We are all soldiers of the revolution. We want the workers' victory over filthy lucre. That is socialism."

In *Mein Kampf*, Hitler wrote that the "folkish philosophy is fundamentally distinguished from the Marxist by reason of the fact that the former recognizes the significance of race...If the National Socialist Movement should fail to understand the fundamental importance of this essential principle... it would really do nothing more than compete with Marxism on its own ground." In a 1923 interview, Hitler declared that communism and Marxism were "not socialism. Socialism is an ancient Aryan, Germanic institution. Our German ancestors held certain lands in common... We are not internationalists. Our socialism is national. We demand the fulfilment of the just claims of the productive classes by the state on the basis of race solidarity."

The Nazi Party platform of 1920, which Hitler helped draft, listed among its demands "the nationalization of all businesses which have been formed into corporations [trusts]," "profit-sharing in large industrial enterprises," "the extensive development of insurance for old age," "the immediate communalizing of big department stores, and their lease at a cheap rate to small traders," and "the passing of a law for the expropriation of land for communal purposes without compensation; the abolition of ground rent, and the prohibition of all speculation in land." (Hitler later "clarified" that, since the party upheld private property, "confiscation without compensation" referred "merely to the creation of possible legal means of confiscating when necessary, land illegally acquired, or not administered in accordance with the national welfare. It is therefore directed in the first instance against the Jewish companies which speculate in land.")

As the above makes clear, Nazis were socialists—who were also racists.

Today's socialists often claim that the racism of the Nazis disqualifies them as socialists. That is a very peculiar claim, given the fact that many of today's socialists are obsessed by race and use the supposed racism of their enemies—an imputation often itself racist—as a rallying cry. For them, race is a vital means of gaining power. Having an enemy to focus on—for example, racists or alleged racists—is vital to mass movements like socialism. In *The True Believer*, his classic study of mass movements, Eric Hoffer points out that when "Hitler was asked whether he thought the Jew must be destroyed, he answered: 'No...We should have then to invent him. It is essential to have a tangible enemy, not merely an abstract one.'"

Moreover, if Stalin, Mao, and Hitler were not "real" socialists, why did they say they were? Simply to confuse their followers? Although we cannot get into the heads of these three sociopaths, we do know they considered themselves socialists. The definitions of communism and socialism specify pretty much the same thing; most crucially, they refer to a social system in which most property is owned or controlled by the state. Communism spins a more elaborate theory to justify itself than does ordinary socialism. For example, communism envisions a class struggle to overthrow capitalism, and there is talk of taking "from each according to his ability" and giving "to each according to his needs." Just as medieval feudalists believed that kings ruled by divine right, so Marxist and allied theories supposedly give the communist leader his right to rule. Many socialists substitute a worship of Karl Marx for worship of God.

Both the Nazis and the Soviets were mass murderers. The communists were first. In the early 1930s, Stalin murdered five million Ukrainians by starvation. The first great mass murder of the Second World War was the Katyn Forest massacre in early 1940, when Stalin had thousands of Polish military officers killed. The governments of Stalin and Hitler both had concentration camps. The Soviet gulags lasted from 1919 to 1987, the Nazi Konzentrationslager from 1934 to 1945. Both the Nazis and communists murdered millions of camp prisoners, but many more millions of prisoners were incarcerated in Soviet camps than in Nazi camps. The murders that Chairman Mao committed for socialism totaled 60 to 70 million. The prospect of such vast slaughter did not dismay Marx. In 1848, he asserted in an article for *Neue Rheinische Zeitung* that the violence of counterrevolution had shown "that there is only one way in which the murderous death agonies of the old society and the bloody birth throes of the new society can be shortened, simplified and concentrated, and that way is revolutionary terror."

Socialists sometimes make the outlandish claim that Nazi Germany was really a capitalist nation. But a capitalist society is an economically free society. In Nazi Germany, company "owners" did no more than follow Nazi government orders, just as the managers in the Soviet Union followed Soviet government orders. Neither could safely defy the demands of a totalitarian government. Along with the Jews and members of other groups, dissenting Germans were sent to Nazi concentration camps; dissenting Russians were sent to the gulag. In both of these one-party states, socialist theory rationalized and socialist practice enabled a country-wide chamber of horrors

Why does Pope Francis encourage socialism? Is the political thought of Pope Francis, born Jorge Mario Bergoglio, consistent with the moral precepts of Jesus of Nazareth? If you examine the pope's words and deeds, including his dealings with the Chinese Communist Party and Xi Jinping, the natural conclusion is that he is preaching and practicing the opposite.

Many words have been attributed to Jesus. Near the top of the list of statements that he most likely uttered is "Render unto Caesar the things that are Caesar's, and unto God the things that are God's." In saying this, Jesus drew a bright line between his followers and political leaders. He was saying that his followers must cleanly separate themselves from the powerful unbelievers.

The idea that Jesus might ever turn over the day-to-day leadership of his followers to a political elite is unimaginable. However, this is exactly what Pope Francis did when he gave the right to name Roman Catholic bishops in China to the Chinese Communist Party—a Party that walks in the footsteps of Mao Zedong, a homicidal sociopath.

Try to imagine how Pope John Paul II—who encouraged Roman Catholics enslaved by the Soviet Union to resist that enslavement—would react to this act of betrayal. Willingness to commit it suggests that Jorge Mario Bergoglio was somehow warped by his experiences. It seems unlikely that his education, whether formal or informal, helped prepare him to do what he has done.

Although hardly following in the footsteps of Pope John Paul II, Bergoglio is somewhat following in the footsteps of Pope Pius XII. While still the Cardinal Secretary of State, Eugenio Pacelli encouraged the German Centre Party, which represented Roman Catholics, to support the Enabling Act of March 1933 to give Adolf Hitler dictatorial power. On behalf of Pope Pius XI, Cardinal Secretary of State Pacelli was also the signatory of the *Reichskonkordat*, the 1933 treaty of mutual support between an emerging Nazi regime and the Vatican that helped legitimize that regime.

Jorge Mario Bergoglio grew up in Argentina under the influence of Peronism, a hodgepodge of conflicting beliefs the key feature of which is authoritarianism. Authoritarianism is alive and well in the Roman Catholic Church; the pope resembles a feudal king and the priests and bishops are his aristocracy. In return for their fealty, the pope provides unquestioning protection to priests and bishops. The unending sexual-abuse scandals roiling the church have barely affected the support that Pope Francis gives these aristocratic offenders.

Like his Chinese Communist collaborators, Bergoglio exhibits distaste for Western values. Consider his "Document on Human Fraternity for World Peace and Living Together," a joint statement signed by Pope Francis and Sheikh Ahmed el-Tayeb, Grand Imam of Al-Azhar, on February 4, 2019, which declares that "among the most important causes of the crises of the modern world are a desensitized human conscience, a distancing from religious values and a *prevailing individualism accompanied by materialistic philosophies* [emphasis added] that deify the human person and introduce worldly and material values in place of supreme and transcendental principles."

Bergoglio could hardly make it any clearer that he despises the Western values of individual freedom and capitalism. For the Muslim leader to join in condemning Western values merely confirms Islam's core beliefs. Both Bergoglio and el-Tayeb are in fact feudal leaders, who recognize that Western values are the anthesis of feudalism.

When socialists claim that workers in capitalist societies are no different from people in feudal societies and that capitalist bosses hold the same positions that aristocrats held in feudal societies, they are uttering nonsense. Feudal aristocrats swore allegiance to their king. Any company president demanding that managers swear an oath of allegiance to him would be fired by the company board of directors. If the very idea sounds crazy, it's because feudalism and capitalism are so dissimilar.

We're not talking about abiding by terms of an employment contract, which typically enable an employee to quit at any time. Consider this example of a feudal oath (quoted in Fordham University's Internet Medieval Source Book): "By the Lord before whom this sanctuary is holy, I will to [name of feudal lord] be true and faithful and love all which he loves and shun all which he shuns, according to the laws of God and the order of the world. Nor will I ever with will or action, through word or deed, do anything which is unpleasing to him, on condition that he will hold to me as I shall deserve it, and that he will perform everything as it was in our agreement when I submitted myself to him and chose his will."

Imagine workers in a capitalist enterprise swearing such an oath of allegiance to their bosses. The idea is absurd because such oaths embodied the whole spirit of the feudal system, a spirit alien to our system of free enterprise (or even semi-free enterprise). The oaths carried the full force of law because, under feudalism, feudal rulers *were* the law.

An Intellectual Vacuum

Socialists never point to their real-world successes to make the case for socialism. They have no real-world successes to point to. So they seek to win converts by spinning increasingly elaborate and detached theories, substituting intimidating but empty verbiage for substance. (As Gertrude Stein might put it, "There is no there there.") Socialists imply, without ever quite saying it, that first you must believe in order that you may understand. Only then can you receive the promised insight and salvation. Apparently without embarrassment, Louis Althusser declares in his book *Reading Capital* that it is "not possible to read *Capital* properly without the help of Marxist philosophy, which must itself be read, and simultaneously, in *Capital* itself."

One of the most common forms of socialist attack on capitalism is name-calling. For example, socialists sling around the word "bourgeois," a catchall term of abuse that can refer to anything that is evil (or claimed to be evil) in human society. A related tactic consists in "proving," or anyway assuming, that capitalism is grounded in deception. According to this argument, many of us in capitalist societies suffer from "false consciousness," the result of brainwashing by material, ideological, and institutional processes. These processes allegedly obscure the true relations between classes, the true interests of class members, and the true state of affairs regarding the exploitation of the proletariat and other downtrodden groups. In other words, the benefits of a capitalism blind the workers to the fact that they are being screwed by capitalism.

The notion of false consciousness is a way of saying that the people are too foolish to know what's good for them; ergo, socialists must protect people from one another and from themselves. This "protection" is imposed in endless ways, ranging from the ridiculous to the dangerous. An example of the former is a warning on a tube of watercolor paint advising easily confused artists that the pigment "contains a chemical known to the state of California to cause cancer," as if a painter might otherwise use it as salad dressing. An example of the latter is the systematic trashing of the Bill of Rights, perpetrated because socialists believe that people cannot be trusted to exercise those rights sensibly.

The charge of false consciousness can be deployed to undermine any successful and functioning idea, economic system, or government. It is a scam repeated endlessly by socialist intellectuals, who target unsuspecting students in the classroom and churn out vast quantities of convoluted and meaningless prose. Socialists and progressives are always certain that their latest "creative" ideas will save mankind. Eager to create their utopia right *now*, they value immediate, direct action over slower, deliberative processes, and they will tolerate no dissent. In 1970, Louis Kampf, president of the Modern Language Association, explained the proper response to a play not aligned with progressive ideals. "The movement should have harassed Lincoln Center from the beginning. Not a performance should go by without disruption. The fountains should be dried with calcium chloride, the statuary pissed on, the walls smeared with shit." Often, the progressive armory seems to consist of only main two weapons, bloviation and thuggery.

Facts are no inconvenience for progressive assumptions. If a critic suggests that those assumptions fly in the face of reality, in response the progressive need merely define that reality out of existence.

This evasiveness is essential to the contemporary transgender movement. Not so long ago, individuals who refused to accept their sexual nature were considered to have a serious problem. It was likened to persons believing that they were Abraham Lincoln or Jesus Christ. Since it was clear that they were not in fact any other person but themselves, the problem in such cases was self-evident. Today, progressives define away the fact that transgender individuals—men who believe themselves to "really" be women and vice versa—have a serious problem.

Despite the death of the Soviet Union, the socialist zombie is still wreaking havoc. Even as Venezuela is being destroyed by socialism, Americans take seriously an overt socialist like Bernie Sanders and hope that his agenda will be implemented. The reality of past failures does not give socialists pause. So what, they say, if communism and Nazism have failed? The Nazis, Soviets, and Chavistas simply didn't do socialism right.

Against self-expression and self-protection. Socialists assail the First Amendment by declaring that so-called hate speech is a crime. There is no limit to what speech socialists may call hate speech, of course. In practice, they justify violations of freedom of speech by claiming that the speech to be repressed would grievously "hurt" some people if not prohibited. The number of people that socialists believe will be "hurt" by hate speech is endless. Government bureaucrats are people, for example; since criticizing the government may "hurt" government bureaucrats, criticism of the government must be banned. The sensibilities of those offended by attacks on capitalism and freedom of speech are mysteriously excluded from the socialists' alleged concern to avoid "hurting" anybody.

Socialists also hate the Second Amendment, which simply and directly states that "the right of the people to keep and bear arms shall not be infringed." Socialists tell you that if you only interpret these words properly, you'll see that they mean the opposite of what they say. The first clauses of the amendment, they stress, refer to "A well regulated Militia, being necessary to the security of a free State..." Allegedly, because the value of a militia is mentioned, the right that shall not be infringed belongs not to the people, as stated, but only to members of the National Guards of the states. Despite the explicit reference to "the right of the people," "the people" is not what the Founders meant. The drafters of the Bill of Rights just didn't understand what they were saying.

This argument is bizarre, especially considering the reality of the United States at the time the Constitution was written. Police forces were rudimentary at best, and would not exist in many states for decades. Much of the United States was a frontier area, at risk of Indian attacks. The notion that individuals had no need for guns to protect themselves and their families would have seemed insane to the drafters of the Constitution. Yet today's socialists deny that individual self-defense, and the right to it, has anything to do with the Second Amendment.

The first American gun control laws were devised in the old South to prevent blacks from carrying or owning firearms. The laws were not enforced against whites. In the decades before the Civil War, restrictive measures were motivated by a fear of rebellion by black slaves that might be abetted if freed blacks could freely own guns. After the Nat Turner rebellion of 1831—in which several dozen whites were killed before it was suppressed—Maryland and Virginia prohibited free blacks from carrying guns, and Georgia prohibited blacks from owning guns. In the infamous *Dred Scott* decision of 1856, Supreme Court Chief Justice Roger Taney stated that one of the grievous consequences of regarding freed blacks as citizens is that they would have the right "to keep and carry arms wherever they went."

After the Civil War, the so-called Black Codes of the south sought to repress the former slaves by, among other measures, prohibiting blacks from carrying or owning guns. According to the Black Code in Mississippi: "no freedman, free Negro or mulatto, not in the military... and not licensed so to do by the board of police of his or her county, shall keep or carry firearms of any kind, or any ammunition." All such efforts contradicted the constitutional amendments passed after the Civil War to protect the rights of blacks (or the rights of all citizens including blacks). One motive for these provisions of the Codes is that letting blacks be armed made it more dangerous for the Ku Klux Klan to terrorize blacks. According David Kopel, author of *The Truth About*

Gun Control, the Ku Klux Klan "was America's first gun control group as well as America's first domestic terrorist organization." Members of the KKK and other white southerners intent upon disarming blacks were thus the forefathers of today's liberal and socialist advocates of gun control.

Socialist distrust of the people also helps explain today's endless government regulations. Socialists don't trust us to run our lives and businesses without orders from the state on how to do it. Some people are not even allowed to work unless they first obtain a government license; nor is this requirement restricted to a few arcane or dangerous professions. Licensing can cover everyone from accountants to hair braiders to welders. Without the intervention of the state, people are supposedly too dull-witted to know whether they like the way their hair is being braided or to patronize an alternative if they don't like it.

Socialists see themselves as a part of a ruling aristocracy entitled to share in the absolute and ultimate power belonging to a king. They see themselves as certainly more worthy than us common folk, from whom sovereignty over our own lives is to be stripped.

Against nature. We are not all equal. Hardly anything in our real world is ever equal. But socialists ignore that fact and fight for what they call "equality."

Half the people of the world are profoundly different from the other half. One half, women, can bear, deliver, and nurse children; men cannot. Neither women nor men five and a half feet tall can play basketball in the National Basketball Association. Anyone can fantasize about making women equal to men, and laws might be enacted to compel NBA teams to keep women on the court half the time. But nobody can argue that once this edict has been imposed, NBA basketball would still be the same game.

Of course, people differ mentally as well as physically. Down's syndrome is tragic precisely because those afflicted have greatly below-average physical and mental capabilities. Nearly all persons suffering from it are incapable of driving a car. Socialists may demand equality for victims of Down's syndrome and may demand special driving tests to make it easier to pass driving tests. But imposing such "equality" would hardly make good sense, as even its loudest proponents often implicitly admit. Despite its commitment to "diversity," for example, Harvard University has never admitted a victim of Down's syndrome into its hallowed halls as a student.

Socialist calls for equality in college admissions are similarly flawed. Socialists do create special tests (or score admission tests differently) to increase the "equality" in college admissions. But since the alleged beneficiaries often cannot do the required coursework, they fail their courses and flunk out of school. The egalitarian "solution" to a natural inequality is obviously no solution, and taking the premise further by granting a college degree to those who have not earned the degree is no solution either.

Socialist calls for equality are another example of disdain for the individual. By lumping people into groups, socialists avoid having to acknowledge individual needs, capabilities, circumstances, and choices or the ethical and social requirements of taking these differences into account. By looking at groups and finding inequalities—inequalities that they often contend, without evidence, are wholly the result of invidious "discrimination"—socialists manufacture endless bogus critiques of capitalism. Since, in their view, members of all groups must meet various socialist-proclaimed goals, fault is easy to find. The contradictions of the approach are evident when socialists demand that 13 percent of Harvard classes be black (in order to more or less match the proportion of the U.S. population who are black), yet fail to demand that no more than 13 percent of NBA basketball players be black.

People have understood for thousands of years, at least since the ancient Greeks, that demands for universal equality are pernicious. The Procrustes of Greek myth who severed the legs of tall people and stretched and broke the bodies of short people to equalize height was not presented as a noble idealist worthy of emulation. The myth dramatizes the folly of attempting to impose equality in defiance of nature.

Against the scientific method. The scientific method is a logical, systematic approach to discovering how things in the universe work. It involves basing one's conclusions and theories about the world—like quantum theory, that vaccines cause autism, or that socialism is the only just and practical social system—on observations of the world, and then testing those theories by reference to further observations, including those made possible by experiments. To the extent that socialists ignore the failures of socialism or ignore relevant facts when coming up with their theory to begin with, they implicitly reject the scientific method, according to which no theory is spared the requirement of real-world derivation and confirmation.

The Chinese government implicitly acknowledged that socialist theory is false when its officials resorted to the incentives of an at least partially capitalist system so that the economy could thrive. China's real-world test of socialist and communist doctrine—a test that proved the inability of socialism in practice to achieve the socialists' own professed goals—has proved too painful for many socialists to acknowledge. So they don't acknowledge it. They simply ignore the reality and continue to claim that socialism is the only way to achieve "fairness and equality."

Socialists indulge in the same contempt for facts in every area with which they concern themselves. When a socialist theory or interpretation is proposed, the socialists brand anyone who challenges it as racist, fascist, capitalist-pig, and so forth. These days, anyone who dares question socialist global warming theology is accused of being a "denier." But, of course, whether denial is a bad thing or a good thing depends on what is being denied and why—and what is being simultaneously affirmed.

Questioning Orthodoxy

When Pierre-Marc-Gaston de Lévis said that it is "easier to judge the mind of a man by his questions than by his answers," he was expressing a key tenet of the Enlightenment, i.e., that every belief must be questioned to determine whether it is supported by fact and reason. When questioning is forbidden, progress stops.

Prior to the Enlightenment and even during its earliest years, before the new science triumphed, danger threatened persons like Giordano Bruno and Galileo Galilei who questioned the orthodox belief that the earth is in the center of the universe. Today's Counter-Enlightenment would return us to the Middle Ages. Socialists believe that uniformity of thought—their thought—is the ideal. In their view, the only thoughts that should be expressed are those reflecting the "progressive" world view. Their shouts of "Denier!" at any who challenge global-warming orthodoxy or "Racist!" at any who challenge Islamic extremism echo the Medieval church shouts of "Blasphemer!"

Such contempt for open and rational discussion is not the province only of goofy fringe groups. For example, in 2017 Google, the giant search firm, fired an engineer named James Damore because he asked a question: whether women may be biologically less suited to perform software engineering jobs than men, so that biology more than any irrational discrimination was more causally relevant in explaining why employees in the tech industry are disproportionately male. He communicated this question to colleagues in a memo entitled "Google's Ideological Echo Chamber," a heinous act for which he was summarily fired by Google CEO Sundar Pachai. Asking such questions makes women feel "judged," "intimidated," and "under threat," Pachai averred. One wonders whether the strong women working at Google feel patronized by this "caring" attitude.

Prohibiting open discussion of a question that is in no way yet definitively resolved is the antithesis of the spirit of the Enlightenment. Such behavior reflects a medieval mindset, and the words of Sundar Pachai echo Savonarola's preaching against Florentine heresies and impieties.

To some persons, always following conventional wisdom may seem the safest course. They may believe that this is the way to avoid ruffling feathers. But the risk of discomfiting others is inherent in being independent and creative—in looking at the world with your own eyes and mind. Being "safe" requires giving up your own creative thoughts and questions, merely plodding along, the better to maintain your place in the herd.

Members of today's academic left treat one research area the same way many people treated the subject of homosexuality a century ago, when it was not a topic of polite conversation and was condemned as a sinful perversion when it was discussed. The equivalent, today, is the question of how genes affect intelligence. It is the rare academic who dares suggest that the relationship between genetics and intelligence is an appropriate focus of polite research. The political scientist Charles A. Murray is called a racist and white supremacist for daring even to mention, let alone investigate, the possibility of such a relationship.

Ironically, Harvard University has unwittingly conducted research that may provide credible data on the matter of race and intelligence.

For many years, Harvard has used arbitrary criteria to discriminate against Asian-American applicants in order to promote "diversity," and the university has now been sued as a result. Scores on the Scholastic Aptitude Test are the most objective single measure of an applicant's qualifications for admittance into universities. If Harvard had chosen to use only these scores to make their judgements, it would have admitted many more Asian Americans than it has admitted and almost none of the African Americans that it has admitted. As reported in the October 22, 2018 issue of *The Harvard Crimson*, the multi-year average of single-section SAT scores, by race, of successful applicants was 767 for Asian Americans, 745 for whites, 718 for Hispanics, and 704 for African Americans. Each racial cohort admitted by Harvard should be a fair sample of each race of the students that Harvard deemed worthy of admittance. It would take only a tiny step from this data to a consideration of the relationship between intelligence and race.

Many scholarly articles show a clear relationship between SAT scores and intelligence, sometimes expressed mathematically as IQ = f (SAT score). Harvard's accidental research could provide the data for a paper that the school itself would never sanction, one that quantifies Harvard's unadmitted conclusions regarding the relationship between race and intelligence.

It is unlikely that student grades at Harvard are also a good measure of intelligence. The grading may well be slanted in many cases because of the need of professors to confirm the rightness of Harvard's admissions process and their own need to defend the egalitarian mantra.

White Privilege, Black Privilege, Yellow Privilege

Defined by lexico.com as the "inherent advantages possessed by a white person on the basis of their race in a society characterized by racial inequality and injustice," the notion of "white privilege" is the most recent construct spawned by socialism's pervasive obsession with race, an obsession that had one of its first explicit expressions in Karl Marx's periodical *Neue Rheinische Zeitung* (*New Rhenish Newspaper*), in which Marx endorsed Engels's suggestion that occasional racial genocide is one way to deal with political problems. In 1849, Engels wrote that in Austria, the Germans and Magyars "will be set free and wreak a bloody revenge on the Slav barbarians. The next world war will result in the disappearance from the face of the earth not only of reactionary classes and dynasties, but also of entire reactionary peoples. And that, too, is a step forward." The fact that that Marx and Engels regarded the Slavs as a "barbaric" (and thus inferior) race that was ripe for genocide is a clear precedent in Marxism for Hitler's death camps.

To understand "white privilege," it helps to examine the "yellow privilege" that Harvard University has sought to counteract when processing its admissions. Harvard officials observed that many Asians were achieving high SAT scores, which, going by objective criteria, went far to earn them admission into Harvard. But these Asians were crowding blacks out of the student body, thereby interfering with Harvard's search for racial "diversity."

Harvard felt compelled to fix the alleged problem. It did so by inventing a set of personal-characteristic ratings, ratings that included things like having a positive personality and being likeable, courageous, kind, and widely respected. Once incorporated into Harvard's new criteria for admission, the personality ratings proved to be an effective means of arbitrarily denying admission to Asian applicants. The new criteria enabled Harvard officials to rationalize the Asian racial quota that the university now established in order to combat what it regarded as yellow privilege.

Is there any "black privilege"? Sure. This corrupt concept can be applied to any racial group. To find it in this case, we need look no further than professional sports.

Consider the National Football League (NFL) and the National Basketball Association (NBA). Players belonging to these professional sports organization are paid incredible salaries. The average annual salary of NFL players is $2.7 million, and that of NBA players is $7.7 million. Although blacks make up only about 13 percent of the U.S. population, black players make up about 70 percent of the NFL and the NBA. Clearly, the great wealth earned by so many black players in the NFL and NBA can only be the result of black privilege—that is, if we adopt the assumptions underlying the venomous notion of "white privilege" and blind ourselves to actually relevant factors.

Socialist obsession with race has produced critical race theory and, with it, the concept of white privilege. Such theories have many modern antecedents. Fear of a "yellow peril" led in 1875 to the Page Act, which banned the immigration of Chinese women; and led in 1882 to the Chinese Exclusion Act, which banned all Chinese immigration.

To deny that Harvard is acting to overcome what it perceives to be "yellow privilege" or "yellow peril" is futile. Harvard was one of the birthplaces of critical race theory, one of the cauldrons that cooked up the invidious idea of white privilege. The Marxist antecedents of the theory are clear in the claim of Ian Haney López, a professor at the University of California, Berkeley, that "capitalism and racism co-evolved under colonialism." Together, critical race theory and its notion of "white privilege" provide Harvard with a rationalization for choosing winners and losers in accordance with race, not merit. The losers may be either Caucasian or Asian.

One of the founders of critical race theory was Derrick Bell, the first black professor of Harvard Law School—appointed as such after socialist students conducted protests to demand that Harvard Law install a black professor. Thanks in large part to his writings and political activism, critical race theory has grown into a thriving cottage industry on the fringes of academia. That this jargonized form of racism now rationalizes Harvard University's racial discrimination against Asians is no surprise.

Obsession with Race

What better way to destroy capitalism than to destroy the foremost citadel of capitalism, the United States of America? This is what we see socialists attempting every day, both intentionally and unintentionally.

The Soviet communists had their useful idiots, and the socialists or neo-Soviets of today have their own useful idiots, i.e., easily manipulated people who wittingly or unwittingly assist socialists in their efforts to destroy America. Football players who refuse to stand for the national anthem at NFL football games are useful idiots, implying by their refusal that the United States does not deserve the respect that standing for the national anthem conveys. Such overt demonstrations are one more step on the road to replacing the U.S. Constitution with socialist rule-by-edict. Why should any reasonable person give allegiance to a nation that does not even deserve this simple show of respect?

The demonstrations are motivated by vague complaints about "bias," but attacks on the country's Founding Fathers are more specific, accusing them of being creatures of their time by owning slaves or—in the case of Thomas Jefferson—possibly fathering children with one of his slaves (although the DNA "proof" of this relationship could with equal or greater plausibility implicate a relative of Jefferson).

If the Founders were so imperfect, the United States must have been badly formed and, therefore, deserve no respect or allegiance. Encouraging refusal of allegiance to a nation is a necessary step in overthrowing it.

Socialists are obsessed with race just as the Nazis were obsessed with race. Nazis hated all "non-Aryan" races and attacked enemies as "Jew lovers." Contemporary socialists routinely damn their enemies as "racist!" and stage theatrical events to "prove" this. Situations involving black or Hispanic individuals become socialist-manipulated dramas exemplifying the racist nature of the enemy. The purpose of the ideological dramas is to de-legitimize America.

The Ferguson riots. The Ferguson, Missouri rioting of August 2014 was a staged drama.

What happened? Michael Brown, a young black man who had ingested marijuana, robbed a convenience store, stealing a box of cigarillos. Police were notified. While driving along in his Chevrolet Tahoe, white police officer Darren Wilson realized that Brown—who was carrying a box of cigarillos and walking in the middle of the street with a friend—matched the description of the suspect in the nearby robbery. The pair ignored the officer's request that they move to the sidewalk. Wilson parked his vehicle and tried to get out. Brown slammed the door and punched him in the face through the open window. When Wilson drew his pistol, Brown tried to grab it and use it against him. In the ensuing struggle, the police officer eventually shot Brown in the arm, who then ran, leaving blood in the car. But Brown ran only a short distance before turning and charging. While backing away from his assailant, Wilson, now out of the car, shot and killed him.

Staging the socialist drama of a racist white cop arbitrarily shooting an unarmed black man required spreading lies about how Brown had died. The truth: when he was shot, Brown was charging toward the officer. The lie: when he was shot, Brown's hands were in the air and he was begging for his life. The physical evidence confirms that the officer was backing away from Brown while shooting, since some of the shell casings ejected from the officer's gun lay *behind* Brown's body. Nor did anything in the autopsy of Brown contradict Wilson's account. An FBI investigation found no evidence that Brown's hands had been raised when he was shot. An investigation by the U.S. Department of Justice also found no wrongdoing by Wilson. But echoing a fiction of what Brown was doing in the moments before he was killed, the rioters who took to the streets of Ferguson to protest what they called murder chanted "Hands up! Don't shoot!"—as prelude to looting and vandalizing.

Two sequels were also staged in Ferguson: the first in November 2014, when the district attorney properly refused to indict Wilson for murder; the second on the first anniversary of the shooting. In each case, the mob's looting and vandalism were attended by disinformational chants of "Hands up! Don't shoot!"

Half-baked standards. Oberlin College, one of the country's most "progressive" institutions of higher learning, actively helped some of its students to defame and harm Gibson's Bakery, a shop near the campus that had a contract with the school to supply baked goods to its dining halls.

An underage Oberlin student who happened to be black tried to steal wine from the bakery. When an employee, Allyn D. Gibson (a son and grandson of the owners), acted to stop the shoplifter, he and two other Oberlin students attacked Gibson. The police were called and the students were arrested. The shoplifter later pled guilty to shoplifting. As part of guilty pleas that kept them out of jail, all three students eventually admitted that Gibson's actions were justified and not racially motivated.

Instead of protesting against their fellow students for failing to respect the property rights of a local business and the person of its employee, Oberlin students began protesting against Gibson's Bakery for acting to foil a thief. Oberlin College officials, including Dean of Students Meredith Raimondo, helped the students conduct the protests by supplying them with gloves and pizza, passing out flyers, etc. The school also suspended its contract with Gibson's—not because the bakery was failing to supply baked goods as contractually required, but because the college wanted to "de-escalate" the situation by appeasing the protesters. The college also said that it would consider resuming its business relationship if Gibson's agreed not to press charges against first-time shoplifters. In a legal filing in the case, Oberlin asserted that the bakery's "archaic chase-and-detain policy regarding suspected shoplifters was the catalyst for the protests. The guilt or innocence of the students is irrelevant to both the root cause of the protests and this litigation."

The protestors called Gibson's Bakery "racist" and claimed that it had a long history of racism. No evidence was provided, and two black employees of the bakery testified that the charge was baseless. "Never, not even a hint [of racist treatment of employees]," said Clarence James, employed by the bakery since 2013. "Zero reason to believe, zero evidence of that...When you steal from the store, it doesn't matter what color you are. You can be purple, blue, green; if you steal, you get caught, you get arrested." *Rolling Stone* reported that "according to police records, 80 percent of those arrested for shoplifting at Gibson's from 2011 to 2016 were white, while 15 percent were black, a breakdown that is relatively consistent with the actual demographic makeup of the city." (Oberlin would try and fail to get these police records excluded from the trial to determine the fate of Gibson's lawsuit against the college.)

Unconcerned with facts, and abetted by Oberlin, the protesters virtually shut the bakery down, greatly harming its earnings and reputation. The fact that the Gibson family is white and the shoplifter is black and the fact that the shoplifter was apprehended by a Gibson family member was enough to fuel this instance of progressive outrage and convict the Gibson family of "racism."

The Gibson family did achieve some justice, however. After it sued the school, a jury awarded it $44 million in damages, an amount that included $33 million in punitive damages. A judge later reduced the total compensatory and punitive damages to $25 million. Oberlin College is appealing.

Duke lacrosse. The 2006 Duke lacrosse case was another staged racial-hate drama. Three white members of the Duke University lacrosse team were falsely accused of raping Crystal Mangum, a black student at North Carolina Central University. The rape supposedly happened at a party hosted by the lacrosse team.

Since the alleged perpetrators were white and the alleged victim was black, in the minds of race-conscious progressives the whites had to be guilty, a conclusion they reached prior to full investigation and examination of evidence. The accusation alone was enough for Duke University, which promptly suspended the lacrosse team, forced its coach to resign, and canceled the rest of the lacrosse season. An ad signed by 88 Duke University faculty members ("What Does a Social Disaster Sound Like?") appeared in the university newspaper to decry allegedly endemic racism on campus perpetrated by whites. "We are listening to our students…These students are shouting and whispering about what happened to this young woman and to themselves." Members of the national news media joined the bandwagon. Here again was proof that modern American society was racist and corrupt. (Months later, another group of professors at Duke issued a statement to distance themselves from the Group of 88 and to explicitly "welcome all members of the lacrosse team, and all student athletes, as we do all our students as fellow members of the Duke community, to the classes we teach and the activities we sponsor.")

After it became clear that Chrystal Mangum had invented the rapes, the national media gave the story of the collapse of her accusations scant attention in comparison to their abundant coverage of the original accusations. The revelation that she had lied and the fact that the entire lacrosse team were innocent failed to support the narrative that America is a failed and racist nation. Again, though, the victims in this case received some measure of justice. Durham County District Attorney Mike Nifong, who in his progressive zeal had withheld exculpatory DNA evidence, was ultimately disbarred, removed from office, and jailed. The accused students sued Duke University and eventually won a multi-million-dollar settlement.

The police often suffer most from the socialist obsession with race. Of course, sometimes police officers do behave badly. But this is not automatically the case. Yet when a white police officer makes an honest mistake involving a nonwhite person, that policeman is likely to be branded a racist who must be drastically punished—regardless of the circumstances and the evidence.

Religion and Socialism

If we cannot deny that Pope Francis governs a more-or-less feudal organization, favors socialism, and is willing to collaborate with the Chinese Communist Party, there is still the larger question of how socialist is Christianity as such. And how socialist is the world's second-largest religion, Islam.

Jesus said "Love your neighbor," defining "neighbor" as everyone else. Islam explains at length how Muslims must fight the enemies of Islam. In the Koran, the sword verses spell out how Islam's enemies must be fought and killed. One can say that Islam is "the religion of peace" only by ignoring the holy writings of Islam. Christianity's holy writings, by contrast, never tell Christians to kill Christ's enemies.

Over the past six centuries, Islam has contributed nothing of much significance to civilization in the realms of art, music, science, engineering, technology, economics, and philosophy. The few contributions that Muslim societies have made are relatively insubstantial and occurred mostly over a thousand years ago.

Musically, Islam is a void. By contrast, music specifically for Christian worship formed the basis for much of Western—and world—music. Many musical instruments were created primarily to play Christian music. The same Islamic void exists in science and technology. These days, the small state of Israel receives more patents than are granted to the entire Islamic world. From 1977 through 2015, the five largest Muslim nations, with a total population of about 800 million, received 758 patents. During the same period, the state of Israel, with a population of about 8 million, received 35,900 patents.

Why have Muslims been left behind? Perhaps Islam's founding documents are one problem. Capitalism has made the advances of modern civilization possible. But the Koran, setting forth a theocratic model of society, cannot abide the individual freedom that capitalism requires. Islam is stuck with the Koran in the same way that socialism is stuck with Marx.

Western nations gave birth to the Renaissance and the Enlightenment. No similar period of sustained intellectual ferment has ever existed under Islam, and the scattered intellectual contributions of a few individuals that historians do mention mostly happened over a thousand years ago. In previous Islamic societies, whenever the occasional individual attempted to create new lines of inquiry that might upset the status quo, sooner or later those attempts ended in failure. Muslims risked death if they questioned Islam itself or whatever Islam supports; such discouragement of questioning also discouraged innovation and initiative. The result was that Islamic nations stagnated, ultimately contributing little to the advance of civilization. These ideas and policies, and their results, continue in modern times. Today, as in the past, Muslims risk death in dominantly Muslim nations if they deviate from Islam-approved thinking and behavior.

Why should the Renaissance and the Enlightenment have emerged in mostly Christian societies? With the Roman Church and its rigid dogma so firmly in control of the spiritual lives of the people, why didn't obedience to tradition triumph over the original thinking during these periods? But the core of Christianity is a belief in the sanctity of the individual. This honoring of the individual seems to have been sufficient to prevail against an often corrupt and self-serving church. The challenge posed by the Reformation also probably helped.

The political systems of contemporary Muslim-majority nations still discourage individual freedom. Even in places where the more extreme Islamist dictates like Sharia law are not formally enforced, flouting these dictates is dangerous. Islam especially harms women; where Islamic law is strictly followed, they are treated as barely human. The more Islamic a nation, the more insistently women are denied an education. Often they are not allowed to work outside the home. By effectively prohibiting any contributions made by intelligent women, Islam condemns half its people to intellectual idleness.

Why is Christianity able to function more or less comfortably in a world run by secular political institutions, whereas Islam cannot? Perhaps because in Christianity's founding document, the New Testament,

Jesus says: “Render therefore unto Caesar the things which are Caesar’s; and unto God the things that are God’s.” This statement draws a bright line between religion and the state, a line that Islam rejects. The Koran teaches believers to render everything unto Allah.

The New Testament suggests that God will not micromanage mankind and that human beings have free will, i.e., have some control over their own destiny. Christian theology has no place for a theocratic state that controls and directs all actions of all citizens. Christians sense that their life is theirs to live, either to succeed or to fail. So they are comfortable with freedom of thought and speech for all people. Many Muslims cannot accept such freedom, especially the freedom to utter words that challenge Islamic beliefs. This intolerance can have deadly consequences. In 2005, Muslim riots around the world in response to a few irreverent cartoons about the Prophet Muhammad published in the Danish newspaper *Jyllands-Posten* led to at least 200 deaths.

Muslim nations of today, ruled mostly by kings, theocrats, or socialists, are nothing like the secular, liberal democracies of the West. The most populous Muslim nation, Indonesia, is now on the verge of changing from a challenged democracy into a mostly theocratic Islamic state. In Indonesia, the imposition of Sharia law has become a real possibility. That would mean such pleasantries as stoning women to death for adultery, cutting off the hands of thieves, and persecuting those who follow other religions.

Constrained by their religious convictions, Muslims often tend to behave as if they have little belief in themselves, as if they lack the sense of possibilities possessed by Christians. As illustrated by Iraq's Baath socialists, an authoritarian ideology like socialism can be more readily adapted to an Islamic context than capitalism, with its individualist requirements and implications, can be. Socialism is all about state control of the individual, which makes it fundamentally consistent with the theology of Islam.

Although we all know about the fervent Muslims who believe that murdering "infidels" will please Allah, it is also true that many Muslims share the Christian attitude of live and live. The true believers kill such "apostate" Muslims as easily as they kill Jews or Christians. I feel sympathy for most Muslims, relative moderates caught between the murderous true believers and a world suspicious of Islam.

Non-Muslims have no more reason to hate and fear a peaceful Islam than peaceful Muslims have to hate and fear a peaceful Judaism, Christianity, or atheism. But we should demand that Muslims reject acts of mass murder carried out by radical Islam. We should demand that they condemn, individually, any Muslim who commits or supports those acts. Muslims must understand that if they will not condemn individuals who commit such acts of violence, they become complicit in those acts.

It is tempting to conclude that all Islam needs is its own Islamic Reformation. The Protestant Reformation inspired many Christians to withdraw their allegiance from a corrupt religious hierarchy, and individual freedom of conscience somewhat replaced obedience to authority. This may seem an attractive model for Islam to follow. But the Protestant Reformation led to many years of war, the most recent instance of which was the years of bloodshed that wracked Northern Ireland a few decades ago. Religious civil wars are especially nasty, even if they are only an excuse for settling other grievances.

Given the long history of warfare between Sunni Muslims and Shia Muslims, a bloodless Islamic Reformation seems unlikely. The Sunni Muslims of Pakistan now possess nuclear weapons, and the Shia Muslims of Iran are working frantically to develop them as well. Obama's bizarre nuclear deal with Iran may have made a happy outcome even less likely.

The United States was formed, in part, on the basis of religious tolerance. Islam rejects religious tolerance. This irreconcilable difference—one that seems to be widening— has often been ignored. Some Muslim groups wish to replace American law with Sharia law. If a citizen is governed by Sharia law, he is no longer governed by our Constitution.

All American congressmen take an oath defend the Constitution. Do they defend it? Why hasn't Congress passed laws to immediately deport any resident alien who by word or deed works to throw out the U.S. Constitution and replace it with a foreign set of laws like Sharia; to immediately strip naturalized citizens of their citizenship and deport them, if they do the same; and to immediately deny the advantageous tax benefits of 501(c)3 status to organizations who work to throw out the Constitution?

Such legislation would place this issue on the table so starkly that it could no longer be ignored. The proposals would be controversial. But the Constitution is not a suicide pact, and it is worth vigorously preserving and defending.

Is my criticism of Islam "hate speech"? I do not hate Muslims. I do pity Muslims, and I deplore the theology of Islam. I do regret that some might be offended by my words, which I believe are fair and reasoned. I also believe that to shut down any speech—merely because someone is offended—is itself offensive and loathsome. Anyone can claim to be offended by anything. (Speech seriously calling for physically harming or killing innocents is clearly a different matter; one is then threatening and working to bring about assault or murder, not just nebulously "offending" others.)

Recent efforts to prohibit or penalize so-called hate speech is another example of Counter-Enlightenment thinking. The right of free speech is a consistent expression of the Enlightenment demand that the search for truth be unfettered. Freedom of speech is embodied in that jewel of the Enlightenment, the United States Constitution, specifically in its First Amendment. In the

2017 case *Matal v. Tam*, the Supreme Court unanimously held that there is no "hate speech" exception to the First Amendment. As Anthony Kennedy wrote in his separate opinion, "A law that can be directed against speech found offensive to some portion of the public can be turned against minority and dissenting views to the detriment of all. The First Amendment does not entrust that power to the government's benevolence. Instead, our reliance must be on the substantial safeguards of free and open discussion in a democratic society."

FOUR
SOCIALIST DOOMSAYING AND GLOBAL WARMING

The religious nature of the socialist movement is manifest in its ecstatic embrace of the global warming theology—lately slightly repackaged as a concern about "climate change"—according to which the climatic patterns of the earth are hurtling toward a climate catastrophe. Every instance of bad weather is adduced as proof of this claim, and the villain in the drama is mankind and our wicked industrial ways. Whether the earth burns to a crisp or freezes into an ice cube, it's the efforts of human beings to survive and prosper that are said to be responsible.

Global warming theology is the perfect tool for smearing capitalists and enlisting impassioned soldiers in the battle to destroy capitalism and enshrine the socialist faith. It offers the ultimate condemnation of capitalism, that it will destroy the world. It offers a rallying call to potential converts: "Help us save the world!" It endows believers with instant virtue: "You are helping to save the world!" It permits a sweeping dismissal of every glorious capitalist success since the spinning jenny, since, after all, it's the smoke stacks and carburetors of industrial civilization that have led to the evil emissions of carbon dioxide; had we stuck to hunting and gathering, we wouldn't be in this mess. And it provides the key to salvation: "going green," i.e., becoming socialist and poor. The end is nigh, and we must repent.

How Marx Mimics Paul

Whether they know it or not, socialist devotees of global warming theology are emulating the apostle Paul, who as history's greatest franchisor helped develop the message of Jesus of Nazareth into the world's largest religion. (His first letter to the Corinthians consists of instructions to the franchise in Corinth, Greece on how to improve the quality control of the operation.) One of Paul's early messages was that the world is ending soon, and he told potential prospects that they should become Christians if they wanted to save themselves. Such a recruitment line can be very persuasive, unless and until the world refuses to end on schedule. Because the world did not end and there was no second coming of Jesus in Paul's lifetime, the end-of-the-world message receded in his later writings. But this did not spell the end of warnings about the end of the world or "end of times."

Such warnings, frequent in writings from 100 BC to 400 AD, are remarkably similar to many of today's warnings about climate change. Today the climate change alarmists tell us that forest fires, droughts, floods, superstorms, epidemics, and other calamities will result from our burning of fossil fuels. Because of global warming, the air will become unfit to breathe and the oceans will die. Burning fossil fuels is a heinous sin, a sowing of the wind that shall reap the whirlwind.

Consider two early Christian writings, the Book of Revelation, the last book of the New Testament; and the Apocalypse of Peter (or Revelation of Peter), one of the many noncanonical works that did not ultimately make it into the Christian Bible but which was read and studied by early Christians.

The Book of Revelation describes a series of disasters that will befall the earth. The disasters are preceded by various dramatic events: a beast emerging from the sea, the breaking of seven seals, the sounding of seven trumpets. The myriad disasters include polluting of the air with foul odors, waters turning to blood, fire being thrown from the sky to burn the forests, and infecting of the people with pestilence.

The Apocalypse of Peter offers less a grand overview of these events than a survey of the horrors that individual human beings will suffer. Some of the wicked will be hurled into lakes of fire, mire, and gore. Others will be "beaten by evil spirits" and their bowels will be "eaten by restless worms."

The socialist purveyors of apocalypse warn us daily of our impending doom. These warnings echo those of earlier prophets of apocalypse, except that the heralding event in this case is the emission of carbon dioxide into the atmosphere. Yet Paul's problem is also theirs: the world is refusing to end. Already, some of the more extreme predictions of world-ending climate change have failed to destroy us on schedule. For example, it is still snowing in southern England, in spite of predictions reported in 2000 by the UK Independent that because of climate change, within a few years snow in winter would become, as University of Anglia research scientist David Viner put it, "a very rare and exciting event...Children just aren't going to know what snow is." (The newspaper's headline was even more definitive: "Snowfalls now just a thing of the past.")

Instead of a thin old man dressed in a robe and carrying a sign proclaiming that the end of the world is at hand, we now have news anchors and climate scientists serving as doomsayers. There have always been people who spent their time by prophesying doom, some of whom even made a living at it. The doomsayers of today are more highly paid than those of yesteryear. But the basic message is the same. We must stop sinning lest we all die.

A Truly Catastrophic Climate Change

The green Antarctic forest teemed with life,
Diverse as any forest ever seen.
From pole to pole, the planet earth was rife
With life forms both terrestrial and marine.

No polar deserts, no bare tundra there.
No grinding glaciers marching from the pole.
A greenhouse earth with climate not severe—
Until *Azolla*'s curse took full control.

The Arctic Sea's *Azolla* bloom did strip
Most carbon from our planet's atmosphere.
And make the whole earth's climate do a flip
Into the icehouse earth we live with here.

Before new ice sheets make our planet worse,
What can we do that will reverse this curse?[1]

Levels of carbon dioxide in the atmosphere were once much higher than they are today, and the earth was warmer. Scientists have found evidence for a so-called *Azolla* event that occurred about 50 million years ago, causing a warm greenhouse earth—warm also at the poles—to become the icehouse earth that we know today. *Azolla* is a freshwater fern that can pull especially large amounts of carbon dioxide from the atmosphere. When conditions are favorable, this fern can grow very rapidly; it can *double* its biomass in two days. The theory is that *Azolla* blooms occurring over the course of some 800,000 years were enough, by themselves, to cause an 80 percent drop in the carbon dioxide in the atmosphere. This in turn induced a gradual long-term cooling of the earth's surface, and finally, perhaps about 15 million

[1] Why write an Elizabethan sonnet to make a point? Others have, so why not?

years ago, freezing of the north and south poles.

A certain congruence caused this change: a more or less isolated Arctic sea and the presence of *Azolla,* a tough plant that thrives under certain conditions, becoming a larger and larger biomass. That such a mass of plant life seems to have stripped most of the carbon dioxide from the atmosphere and radically changed the earth's climate is amazing. We are forced to suspect that these ferns failed to heed the strictures of climate change orthodoxy.

This orthodoxy demands that the level of carbon dioxide in the atmosphere continue substantially unchanged with respect to an arbitrarily selected baseline. Since the proponents of this orthodoxy live in the present, what they contend is that our present-day icehouse earth climate must not change. It is our present-day climate that is optimal—or perhaps the climate of twenty or thirty or sixty years ago, depending on when man's industrial activity is said to have begun pushing average global temperature in the wrong direction.

Well, why? Why are polar deserts, bare tundra, and glacier-covered landscapes better than green forests?

What is the worst ultimate fate that we might suffer as a result of rising levels of carbon dioxide in the atmosphere? A return to something like the greenhouse earth that flourished before the *Azolla* event? A greenhouse earth would have no polar deserts, no bare tundra, no glaciers, no fields of ice. Might losing all that polar wasteland constitute a net benefit for mankind? Or would such a boon be more than wiped out by the trauma of rising sea levels and drowning coastal cities?

Sea-level Hysteria

Aren't we raising sea levels by our carbon emissions? Won't this cause catastrophic flooding? Should we not take all possible actions to avoid that?

Let's do some traveling. Imagine yourself in Cape Hatteras, North Carolina—part of the barrier islands. Looking eastward, your seashore view is of the broad Atlantic Ocean as far as the eye can see.

Now imagine that—still standing on the same patch of land that will one day be called Cape Hatteras—you have traveled back in time 15,000 years to a moment perhaps 5,000 years after the first human beings made the long journey from Asia to the North American continent. Perhaps you are in the company of several native Americans, combing the hillside for food. Now as you look to the east, there are trees and grasslands as far as the eye can see and no sign of the Atlantic Ocean. Its seashore is about 50 miles to the east.

Now imagine that—still standing on the same patch of land—you have traveled back in time to the height of the previous interglacial period, some 100,000 years ago, long before the ice sheets of our current ice age have intruded. We may call this interglacial period "the caveman interglacial." Humans were around then, but they were what we might call cavemen. Now when you look eastward, you don't see much. You can't even breathe, for you are thirty feet below the ocean's surface. Back then, the sea level was thirty to thirty-five feet higher than it is today. Don't blame the cavemen, though. Their carbon footprint was insignificant. The change in sea level occurred entirely without their assistance.

What happened during the caveman interglacial period will almost certainly happen again, during this, our very own interglacial period. It isn't reasonable to suppose that we can prevent a similar rise in sea level, notwithstanding any histrionic claims that human activity will flood the world. If you mention this previous rise in sea level to a climate-change true believer, the response will be that the increase in atmospheric carbon dioxide caused by our industrial activity will cause flooding to occur sooner than it otherwise would have. But he will offer no proof of this claim.

Intemperate tropics. Socialist doomsayers claim not only that human-induced global warming will raise sea levels but also that this warming will make tropical regions too hot to support life.

Fifty million years ago, the earth's atmosphere contained much higher levels of carbon dioxide than it does today—perhaps ten times higher. Well, if inducing comparably high levels of carbon dioxide today would render large portions of the earth uninhabitable, then such levels in ancient times must also have rendered the tropical regions uninhabitable back then.

However, we know that the ancient earth, including in the tropics, hosted a wide variety of life. According to impending-doom theory, the ancient earth's only habitable regions would have had to be regions of high latitude. But no reputable scientist has ever claimed that life on ancient earth was impossible in tropical regions. Ancient life existed over the entire earth, including the tropical regions, which were not "too hot to support life." So there is no reason to believe that capitalism-induced increases in levels of atmospheric carbon dioxide would render the tropics uninhabitable in our own day. We must keep in mind, also, that human beings have always been able to adapt to extremely different climates. We live in Alaska as well as Brazil, and in many parts of the world, the very same persons routinely manage to cope with both hot summers and cold winters.

Tilted Windmills

> Wind power's a fraud, don't you know.
> 'Cause a subsidy's what makes it go.
> Greens cry, "Give us your trust!"
> But their program's a bust,
> 'Cause so often, the wind doesn't blow.

Climate change zealots claim that renewable or "green" energy will "save the earth" by maintaining the current level of carbon dioxide in the atmosphere and thereby (presumably) maintaining our present climate.

The unreliability of wind power. The John Muir Trust investigated whether intrusive and unsightly wind farms are delivering power in a way consistent with what proponents have claimed: i.e., that wind turbines will, on average, generate at least 30 percent of their capacity over the course of a year; that periods of widespread low wind are rare; that it is very unlikely that periods of very low wind output will coincide with periods of peak demand for electricity; and that if and when there are prolonged periods of low wind, pumped-storage hydroelectricity can fill the gap. Claims like these are used to justify lavish taxpayer-funded subsidies of wind power.

A study by Stuart Young Consulting for The Trust, *Analysis of UK Wind Power Generation: November 2008 to December 2010*, investigated whether these claims are factual. The researchers measured the power actually delivered to the British power grid by all wind farms in Scotland from November 2008 to December 2010. They found that efforts to harness wind do not in fact provide enough power to justify the lavish taxpayer-funded subsidies. (Of course, if they did, what would be the need for subsidies? If wind power were economically feasible, it could form the basis of profit-making enterprises, just like every other feasible means of satisfying consumer demand.)

The researchers learned that:

- During these 26 months, there were 124 low-wind periods during which power output averaged only 4 percent of rated capacity. Not nothing, but close.

Low-wind periods lasted almost 5 hours on average and, on average, occurred every 6.38 days.

- During each of the four periods of highest peak demand in 2010, the contribution of wind power to the power grid was pitifully low: 4.72 percent, 5.51 percent, 2.59 percent, and 2.51 percent of the output capacity of the wind farms. Wind power just wasn't available when power was most urgently needed.
- The claim that wind turbines would deliver an average of 30 percent of their rated capacity over the course of a year is false. Wind power provided only 24.15 percent of rated capacity. Even apart from the other deficiencies of wind power, this 20 percent shortfall from an already-low 30 percent of capacity must severely impair the economic viability of wind farms.
- Since pumped-storage hydroelectricity provides only a few percent of the total capacity to generate electrical power in the UK, it can "fill the generation gap during prolonged low wind periods" only as long as wind power provides only a few percent of total capacity to generate electrical power in the UK.

The study also found that during the period reviewed, wind power was generated at less than 20 percent of rated capacity more than half the time, less than 10 percent more than a third of the time, less than 2.5 percent for the equivalent of one day in twelve, and less than 1.25 percent for the equivalent of almost a day a month.

How cheap is wind power? Radical environmentalists often claim that renewable power will soon be cheaper than power produced by fossil fuel. To "prove" this assertion, they not only assume that the false claims refuted above are true but also assume that all electrical power is the same. Their badly flawed method of comparing different power generation technologies, levelized cost of energy, makes that assumption explicitly.

But all electric power is not the same. In the cradle of renewable energy, California, the price of electrical power offered to residential customers varies from 13 cents per kilowatt-hour to 44 cents per kilowatt-hour. Electric power produced and used in California changes in price by a factor of three. The price of electricity depends in part upon when you use it. At times of high demand, the price of power is higher than at times of low demand.

Using levelized cost of energy is nonsense when comparing different power generation technologies. It is the wrong tool for comparing different technologies for generating electrical power.

When supporters of wind power use levelized cost of energy to "prove" that wind energy is competitive with fossil energy with respect to producing electricity, they ignore the fact that wind power is unavailable when you need it. The kind of power that you can call upon when you need it, i.e., dispatchable power, is provided by power plants using fossil fuels and nuclear fuel. Dispatchable power is much more valuable than wind power—or any other kind of come-and-go power—precisely because you can count on it. Using levelized cost of energy works as long as each of the technologies being analyzed produces the same sort of power. It is the wrong tool for comparing a technology that produces dispatchable power with a technology that produces a non-dispatchable power like wind power. The concept of levelized cost of energy presupposes that all of the forms of power being compared are the same.

(For a detailed comparison of dispatchable power and intermittent power, see "Comparing the Costs of Intermittent and Dispatchable Electricity Generating Technologies" by Paul L. Joskow, available at bit.ly/329KVC9 through the SemanticsScholar.org site.)

Although the wind blows a bit more reliably offshore, offshore wind power does not solve the problem, since offshore wind power is two or three times more expensive to build than onshore wind power.

To be sure, if costs were irrelevant and consistency in generating electricity were irrelevant, the case for wind power would be a slam dunk.

Politics, Not Science

The world's largest fossil fuel company, Exxon-Mobil, is being sued by governments and organizations because of its massive oil and gas production and sales and because burning oil and gas causes carbon emissions. One popular if flimsy legal claim is that Exxon has not adequately revealed to investors the "fact" that its business model is at great risk given the possibility that future governmental regulations will prohibit the production and sale of oil and gas (and with it, presumably, all industrial activity powered by gas and oil). In response, Exxon has agreed to publicly provide information detailing the business threats of future "carbon dioxide pollution regulations." (Exxon has not yet "revealed to investors" the fact that if governments began taxing Exxon profits at 95 percent, the company could not survive.)

Regarding the scientific aspects of climate-change claims, the company has said nothing. Exxon's muteness is notable because it otherwise makes a great show of grounding its decisions on science. The corporate website proudly details its Operations Integrity Management System (OIMS), a system of scientific and engineering analysis to identify and quantify risks in Exxon's businesses and to develop strategies to minimize or eliminate them. According to Exxon CEO Darren Woods, OIMS "is a cornerstone of our commitment to managing SSH&E risk [safety, security, health, and environmental risk] and achieving excellence in performance."

Given Exxon's commitment to OIMS, it is curious that the firm has never subjected the risk of carbon emissions to an OIMS analysis. But I believe that although Exxon officers have chosen to deal only with the politics of climate change, they do in fact regard the risks of carbon emissions as a suitable subject of OIMS analysis.

Their fearful reluctance to perform it in this case illustrates how the question of global warming or climate change is now mostly a political issue. The issue has become so politicized that anyone who wishes to objectively examine the relevant science is now disparaged as a "denier." While ignoring all the requirements of the scientific method and open scientific debate, the socialists say that the science is "settled," treating their assertions as a dogma that may not be critically and objectively examined. They seek to shut down any rational analysis of climate change and its alleged consequences and causes. Their declarations—one big lie after another—have more the ring of religion than science.

The strategy of the big lie is most commonly associated with Nazi Germany. The Nazis convinced many Germans that the Jews and officials of the imperial German government had somehow "stabbed the German nation in the back" and thereby caused Germany to lose the First World War. But the Nazis certainly do not have a monopoly on this tactic.

Our new president, Joe Biden, is acting on the basis of the big lie of global warming, according to which a rise in sea level can be stopped by reducing carbon emissions and the tropics will become uninhabitable unless we reduce carbon emissions. Biden's entire climate-change program is an attempt to fix a problem that does not exist. One of his first acts as president was to cancel the Keystone Pipeline, which will cause economic damage to the United States and Canada. His "renewable energy" program will compel massive funding of uneconomic technologies. But the destructive consequences of his pointless actions are invisible to Biden, who has been seduced by global warming fantasies.

The Power of Hatred

Socialism uses hatred to motivate and recruit. Clear enemies give socialists a means of rallying the faithful. "Hatred is the most accessible and comprehensive of all unifying agents," writes Eric Hoffer in *The True Believer*. "It pulls and whirls the individual away from his own self. . . ." In the early Soviet Union, Josef Stalin chose the Kulaks, among others, as the hated ones. The Nazis chose the Jews. One focus of the hatred of contemporary socialists are the "deniers": anyone who voices doubts about the global warming orthodoxy. Contemporary socialists are sure that everyone overlooks their kinship with the Nazis, so they are oblivious of the irony of using such a term to label those who dare dissent from their received dogmas.

FIVE
CREATING ART

Great art thrives under capitalism. Whenever individual freedom increases and people are allowed to work for themselves and express themselves, art thrives. Ancient Greece had relatively primitive politics, but the classical period was a time of greater freedom for many people, a time during which art thrived. Art also flourished in Renaissance Italy, along with capitalism; in Florence, the Medici, successful bankers and dedicated capitalists, loved supporting new and innovative artists like Michelangelo and Leonardo da Vinci. These artists had freedom to work and travel, and they used their freedom to create enduring works of art. Likewise, today, all of the art forms thrive in the relatively free West.

Art and Freedom

In addition to providing artists with the freedom to express themselves, capitalist societies have also generated another great boon: wealth. A wealthy society can support its artists much better than an impoverished, socialist society can. Moreover, the citizens of a wealthy society are better able to appreciate art. If you are hungry, indulging in the arts is not a priority; it's harder to become a connoisseur of the arts if you are always scrambling to survive and have little of the disposable income that is a commonplace in a free and prosperous commonwealth.

Inspired art is the opposite of planned art. Great artists never set forth a Soviet-style five-year plan to detail their long-range artistic intentions. Great art, and most not-so-great art, springs from the artist largely unplanned. (Artists certainly plan how to perform the execution phase of their work, but these activities involve the routine mechanics more than the "art.") Socialists like Hitler and Stalin suppressed individually directed, creative art. They substituted planned government-mandated Nazi "art" and Soviet "socialist realism."

Stalin lavished support on artists who conformed to the dictates of socialist realism, and their pathetic, dreary paintings are the result. That same dreariness afflicts all aspects of communist societies. I visited the East Germany of the 1960s, the Moscow of the 1970s, and the China of the 1970s; all of these communist societies were dreary and pathetic.

The approved art of communist-style socialism and Nazi-style socialism are identically drab because these social systems are identical evils. They destroy all things human at the root.

One of the most ironic incidents in the history of art occurred in 1937, when the Nazis staged a show of so-called degenerate art (*Entartete Kunst*). Hitler wanted to show Germans that capitalistic, "degenerate" art was worthless. Nazis were not amused when the "degenerate" art drew much larger crowds than their displays of officially approved art.

In both Nazi Germany and communist Russia, artists survived by obeying the dictates of socialist realism. Socialist strictures squeeze the life out of a nation's popular literature as well. The same is true of every field, whether artistic or not; practitioners prostitute themselves to survive, relinquishing their independence, creativity, and best judgment.

Perhaps it was the socialist destruction of Russian literature in the twentieth century that inspired the greatest writer to come out of the Soviet Union, Aleksandr Isayevich Solzhenitsyn. His literary life was a continuing response to the evils of socialism, evils that he had directly experienced. Solzhenitsyn wrote some of the most damning critiques ever of the communist system, the novels *The First Circle* and *One Day in the Life of Ivan Denisovich* and the nonfiction account *The Gulag Archipelago*. You could start with *One Day in the Life of Ivan Denisovich*, which is beautiful and short.

The Act of Creation

A person paints, composes, or invents something unique. We see the product. We don't see the process—exactly how the creation is conceived and developed. An art teacher might explain how to paint a watercolor sky: "Pat a moist, French ultramarine background with a dry tissue." The mechanics of painting can thus be made understandable; but this is only mechanics. Creativity is in the mind, not in the hands. Exactly how is a painting composed? Preparing a composition is creation in its purest form, yet how to create a composition remains a mystery.

Creativity is a bit like pornography. I can't exactly define it, but as Supreme Court Justice Potter Stewart said about pornography, "I know it when I see it."

One popular theory about the functioning of the brain says that the human brain has two separate parts, the right brain and the left brain, which specialize in different kinds of thinking. These terms describe differing brain functions and do not necessarily relate to the side of the brain where these functions take place. Some use the terms R-mode and L-mode to stress the fact that this division does not necessarily refer to the right and left hemispheres of the brain.

Left-brain thinking is basically verbal, analytic, and sequential; right-brain thinking is basically visual, perceptual, and global. Left-brain functions are thought to be things like linear thinking, mathematics, and logic. Right-brain functions are thought to include imagination, intuition, and art, which by their nature seem not to be the product of rational, conscious thought. This idea is attractive, since most of us have experienced that creative flash for which we can offer no logical explanation.

In her book *Drawing on the Right Side of the Brain*, Betty Edwards provides a simple test by which you can demonstrate this kind of right-brain/left-brain distinction within your own brain. In the test, you are provided with a line drawing of a man and instructed to copy the figure twice. First, you copy it while looking at the figure right side up. Then you draw it a second time, looking at the figure upside down. You will be amazed. Your second copy, the copy drawn while looking at the figure upside down, will be much more accurate. This experiment has been conducted tens of thousands of times (maybe hundreds of thousands). Try it yourself, and you're likely to find that, like me, looking at the figure upside-down enables you to do a better job.

Betty Edwards explains that since it already "knows" how to draw a man, the left brain (which is usually in charge) doesn't allow the brain to truly observe the figure. The right-side-up copy suffers from the oversimplification required to be fast and efficient. The right brain is not overconfident, however, nor is it in a hurry. The right brain takes the time to look at the upside-down figure. The right brain really *sees* the figure and thus enables a more accurate reproduction.

Many other methods can disengage the left brain. Suppose you are sitting quiet and still in the mouth of a New Mexico canyon, listening to the wind murmur through the juniper trees. Looking at the azure sky and the scarlet and gold cliff face of the mesa while listening to the random sounds of the wind gives the left brain nothing to process. It cannot "know" what is going on. The right brain, on the other hand, does not need to know. The right brain simply absorbs the feedback from all of the senses without making any judgements.

Here is an example of my right brain at work. I wrote "While Sketching" while sitting at the base of that New Mexico mesa, with my left brain mostly out of the loop.

"While Sketching"

Seeking the shade of a juniper tree
While the breeze sighs its breath over me.
It puffs and it blows,
Now tell me who knows,
What else could be nearly so free?

We were at a watercolor painting workshop at the Ghost Ranch, west of Taos, New Mexico. I had taken a folding chair and my painting gear to a spot on the Box Canyon Trail. That perfect day needed only a shady spot where I could take in the surrounding beauty and begin to paint. I had barely begun when I paused as the sound of the breeze in the juniper trees insinuated itself into my consciousness. A bewitching distraction, it was impossible to ignore. Looking down at my first few tentative brushstrokes, I could not force myself to continue. I had to paint the sound of the wind in the juniper trees, but I could paint the wind only with words.

I soon knew what I wanted to write, a limerick, and I had some idea of the rhymes that I might use. You may say to me, "No, that poem was merely the product of left-brain calculation, and the right brain was uninvolved." Not true.

For example, it had never consciously occurred to me to use any alliteration in this poem. I gave no conscious thought to using it. Note the word combinations "seeking… shade" and "breeze… breath." These alliterations were not a matter of conscious intention. Wholly unplanned, they manifest the right brain in action. I believe the alliterations add a great deal to the poem. (You may not care for them; no matter. I would never argue about it, being a strong believer that de gustibus non est disputandum.) Another example is the concluding couplet of my Shakespearean sonnet in chapter four. Shakespearean sonnets must end with a couplet, but mine also has a couplet within the sonnet's last four syllables. Left-brain calculation? No. My left brain may claim it, but it was pure right-brain invention.

But does this right-brain creativity function only when doing artistic things like poetry or music? No. From personal experience, I know that nearly all acts of creation depend upon a triggered right brain. Something as simple as standing in a hot shower can provide such a stimulus. There was a standing joke in our house. I would finish showering and tell my wife Gayle that "I had this great idea!" (an often novel idea dealing with process design in chemical engineering). Gayle would laugh and say, "Again?"

Some people associate the right brain with emotions. Wrong. The right brain is a different, rational part of the brain. It understands emotions better than the left brain can, because the right brain takes a broader, a more comprehensive view of reality than is possible to the left brain. The right brain may be "smarter" than the left brain, since it can solve puzzles of much greater complexity.

A likely example of right-brain creativity and reasoning may be found in the creation of quantum mechanics. Max Planck investigated black-body radiation. His goal was to determine how the intensity of the electromagnetic radiation emitted by a black body depended on the frequency (i.e., color) of the radiation. This problem had proved impossible to solve using classical Newtonian physics. Planck solved it with a right-brain leap of the imagination. He imagined that energy is emitted only in quantized form, i.e., that energy could only be a multiple of some fundamental, elementary unit of energy specific to the frequency of the radiation. This leap of imagination was incompatible with Newtonian physics.

However, Max Planck's left brain never accepted quantized energy. In fact, Max called his creation of quantum theory "an act of despair." It is unfortunate that Planck did not appreciate the distinction between our left brain and right brain. Doing so might have given him some comfort. Perhaps Planck's Christianity would have invited him to consider whether the right brain holds the spark of the divine that some people believe lies deep within every human being.

Poetry

The Ghost Ranch is a special place. It's no accident that the artist Georgia O'Keefe lived most of her life there. Seeing its beauty might bring out the artist in you too.

A Place Called Camposanto

This land's called Piedra Lumbre, named many years ago.
A haunted, dry, high desert place in north New Mexico.
And in it lies the Ghost Ranch, where artists sometimes dwell.
And on that ranch is a special place, whose story I'll now tell.

The cliffs shine high above us against a cloudless sky.
Their red and yellow colors take hold and seize the eye.
It's a place called Camposanto, here in this sacred ground,
Remembered friends of Ghost Ranch lie, 'mid beauty all around.

But was it always peaceful, this place of calm and rest?
Or once did something happen here, which no one else has guessed?
Next to the shining cliffside lies the talus slope below.
A barren, steep debris field. Of sand. Of rock. Aglow.

And at the talus top, next to the cliff face sheer,
Exists an open narrow space. It's flat and rather near.
But no one climbs up to it, and now it's left alone.
Yet was there once a time there, when it became well-known?

***

Four hundred years have passed since their small band rode by.
An even dozen horseman, not much like you & I.
A troop of young explorers, they mostly came from Spain.
They rode here then for glory. For gold. For their own gain.

They traveled through this land and felt somewhat at home,
Because the Spanish countryside was such a place to roam.
Yet this land held a danger like they had never seen.
A people wild and ruthless, and all things in between.

Those red men saw the Spaniards. And watched them move along.
But most of all they wanted those beasts that looked so strong,
That carried all the Spaniards, together with their load
Of strange and shiny objects that rattled as they rode.

The Spaniards knew the red men, though they'd seen none around.
They traveled with great caution, and stopped at each strange sound.

For they had heard of others who did not take such care.
Whose bones now lay, untended, out in the desert air.

Near the shining cliffs, they stopped, just for the night,
Some distance from a flowing stream; and that, they thought, was right.
For these men all were soldiers. They knew the cliffs would give
Enhanced protection from that side and, maybe, help them live.

The light of day had faded. The shining cliffs were dark.
Their campfire blinked and flickered, and then up flew a spark.
The night guards each were watchful and listened carefully,
But nothing moved or whispered, till dawn broke soundlessly.

Just when the birds were singing and cliffs eased into view
Great whoops and shouts erupted, and swishing arrows flew.
The Spaniards all were startled and struggled to react.

For they had not expected so sudden an attack.

But still they joined together and held their ground a while,
Though soon the red men's numbers defeated Spanish guile.
A panicked, frantic scramble up that talus slope
Was all that saved some Spaniards. It was their only hope.

Not all of them survived, that cool and deadly morn,
'Cause six of them lay dying. Ripped. Bleeding and forlorn.
Their blood that soaked that slope, at once was gone away.
Their red blood sinking in the earth, it barely marked that day.

The six survivors lay on that high, narrow ledge.
Sheltered from the arrows' flight by its rough rocky edge.
The red men had not followed. The Spaniards wondered why.
But it was very simple. No red man wished to die.

The red men had no reason to fight again that day.
The Spaniards were as good as dead. There's little more to say.
For they now had no water, and as the sun did bake,
Their bodies sadly, slowly died, an awful toll to take.

Four days was all it took, until the last one died.
And all that time the red men never even tried
To climb the talus slope and fight them man to man
For they were satisfied that victory was at hand.

***

Four hundred years six Spaniards lay, their bones now white as fleece.
White bones baking in the sun. High up and left in peace.
But as we climb this talus slope, we simply hope to tell
What lies in this magic place, in Camposanto's spell.

Ghost Ranch visitors feel the majesty of this land, which is probably what led us there. Our first visit was a drive-through. On our second visit, I tagged along with Gayle as she attended a watercolor workshop. While she painted, I did some hiking and played golf at the Taos Country Club. On our next visit to the ranch, we both participated in the watercolor workshop, which included some painting outdoors. My *en plein air* studio was a spot below the mesa, just off Box Canyon Trail, near Camposanto, a place with memorials to Ghost Ranch supporters who have passed away.

The last time we visited, we returned with the same group that did the workshop. This time, though, the plan was to paint only, skipping the workshop. But I soon realized that I was really there to write. Camposanto was only a short walk away, so I spent time there at the base of the mesa to regain the sense of the place.

Mesas usually have a talus slope at their base: fallen rock and wind-blown sand and dust. It will always lie against the mesa with a slope of about 35 degrees—not too steep to climb, but a challenge. Standing below and gazing upward, I wondered what lay near the top of slope, against the cliff-face. Later, with one of my hiking buddies, I climbed up to see. We quickly realized how tough the climb really was. Had I been in my twenties, I'd have scrambled up in no time. In my fifties it would have been a struggle. But I was in my seventies now, and it didn't seem wise to risk trying the whole climb. So we reluctantly turned around and made our way back mostly by sliding down the slope on our backsides.

Failing to reach the top of the talus slope, we paused at its base, looking upward. "Let's take a few minutes here," I said, settling on a convenient boulder. My vision gradually came into focus. This was another time. I would not be here again for 400 years. Over there a troop of Spanish explorers was setting up camp, pausing in their search for gold. Their search would not end well.

Telling the story of these explorers gave substance to the flow of light and shadow underneath the cliff face. A place of such beauty deserves its own story, whether fact or fiction.

A system for writing poetry. My thinking on the writing of poetry has evolved to the point at which I now have a system that relies only partly on calculation or left-brain work. I roughly outline the story of the poem and come up with words that may fit that outline. I then visit the site rhymezone.com to find rhymes that also fit the outline. Then I stop "trying" to write the poem and let the right brain take over. Eventually, the poem emerges with little conscious effort on my part. What remains for the left brain to do is only a bit of editing.

Perhaps your response is: "Of course. How else would you do it"? Or perhaps you regard the system as weird or absurd. If you are inclined to the latter opinion, I suggest that you consider the Hawthorne effect and the placebo effect, which describe mental behaviors that at some deeper level seem to be related.

The Hawthorne effect pertains to the fact that individuals sometimes quite unconsciously change what they do when they are aware of being observed.

It was first identified by researchers studying worker productivity who realized that the researchers' own acts of observing had motivated the workers to become more productive. When the workers saw that someone cared about their work, they tended to perform better.

A placebo is a non-therapeutic substance used as a control in drug trials. The placebo effect pertains to a person's response to ingesting what he believes to be medicine but which is not medicine. A person takes aspirin for a headache and the headache goes away. A person takes a sugar pill that someone claims is aspirin and the headache goes away. The sugar pill apparently works primarily because the person thinks that it will work, not realizing that it is a placebo. About fifteen years ago, I had a very annoying case of hay fever that sometimes made it nearly impossible to play golf. My doctor prescribed a very expensive nose spray that more or less cured me. I didn't like the idea of being dependent on a prescription drug, and I wondered whether I might use the placebo effect to cure my affliction. So I bought a saline nasal spray and began using it during hay fever season. I was for the most part cured. I still faithfully use the spray, which costs me only a few dollars a year.

Another example of letting the right brain work its magic is playing bridge. My bridge games used to involve left-brain calculation to the point of agony. I counted cards in suits, considered probabilities, etc. I now usually do not bother to calculate exactly how to bid or to play a hand. I simply look at the problem and let the right brain take over. I still do occasional quick counts and probability assessments, but for the most part I let the right brain take over. Playing bridge using the right brain is much more fun and relaxing. There is not the angst-filled need to win. I suspect that my results now are as good as or even better than they were before.

Painting and Remembering

Although I spent time composing poetry while attending the watercolor workshop at the Ghost Ranch, I also did some painting. One of my paintings depicts my Ghost Ranch hiking buddies standing on Box Canyon Trail.

The Ghost Ranch inspires. Enjoying its beauty liberates the right brain and makes all forms of creating easier. Doing this watercolor was great fun, both for the tactile enjoyment of painting and for the way it anchored these memories of place. Sere, high desert has a stark beauty, and the occasional presence of permanent water creates lush microclimates like the one shown here near the creek. Both plant and animal life are abundant. Birds and animals are drawn to the water. Long ago, humans were also drawn here, as evidenced by the ancient Indian petroglyphs that Jenny, Katie, and I happened upon during our hike into the canyon and along its spring-fed streamlet.

When painting, I tend to focus on place. Whether cliffs or trees or streams, waves thudding onto a seashore, or wind sighing in juniper trees, a place large or small will grab and hold my attention. It may be anything from the endless sweep of a seascape or semi-arid landscape to the window of an old church.

Perhaps if I had made painting a career, my work would reflect a broader range of subjects. But I doubt it. Even many professional artists seem to concentrate on a fairly narrow range of subjects. Vermeer devotes himself to confined spaces, and some of the Impressionists are especially drawn to landscapes. What I try to do is capture the feel of a place.

You can see that in Sea Eagle. The lone eagle accentuates the vast emptiness of the ocean and beach below. The eagle may be the main focus of the painting, but the ocean and the beach are the soul of the work.

It is very satisfying to use art to record the places that Gayle and I have visited. We loved hiking on Stafford Beach, which is on the Atlantic Ocean side of Cumberland Island, Georgia; and in 2010 we spent several glorious days there in the second week of October to celebrate Gayle's birthday. (Let me add a plug there for the Greyfield Inn on Cumberland Island. A great place to stay—in fact the only place to stay on the island, unless you like to camp out.) On the lovely day that inspired this painting, we had the 17 miles of the beach all to ourselves but for a small herd of wild horses and the lone bald eagle drifting above. My painting only suggests the splendor of that day.

On another occasion, we were in Ireland before Gayle and I discovered that we each had Irish ancestors. (Eventually, a DNA test would suggest that I am perhaps one-fifth Irish.)

One of the more enchanting spots we visited was Inishmore, one of the Aran Isles that lie off Ireland's western coast on the opposite side of the Atlantic Ocean from Cumberland Island. Inishmore has always been sparsely populated. Millennia ago, the island's first hardy settlers realized that the island had so little soil that it would be nearly impossible to grow the crops they needed to survive. So some clever capitalists decided to pile up mixtures of sand and seaweed; once the seaweed composted, the resulting soil made it possible to grow crops.

Gayle and I took the ferry to Inishmore and stayed for the day. This poem is my small effort to capture its untamed look and feel.

Aran Isle Day Trip

Rocky hillsides, rocky fences,
Rocky streams where sunlight glances.
Aran Isles wear seascape light.
Summer now, not winter's night.

Waters calm, our leeward side.
But waves crash under cliffs that hide
Snowy birds and tiny flowers.
We sail away. We've no more hours.

SIX
CREATING TECHNOLOGY

All technology flows from the inventions of individuals and, like art, flourishes best in the context of a free society. The desire to improve one's own life and the lives of one's family—as well as love of creation—is what inspires a person to invent.

I loved being a chemical engineer because it gave me so many opportunities to do creative work—specifically, the work of process design.

Consider an oil refinery. A corporation plans to buy crude oil and turn it into gasoline, diesel fuel, and other petroleum products in order to sell them and make a profit. The process design engineers are called in. "We want to build a refinery near Houston, Texas to make X barrels per day of gasoline for sale," they may be told. "Permian basin crude oil will be delivered to the refinery by pipeline. Design a refinery."

Process design starts with a blank sheet of paper. The finished design specifies things like the size and number of storage tanks for the crude oil and for the gasoline, diesel, or other petroleum products; how large the crude oil storage tanks must be if the refinery is to continue operating when the pipeline goes out of service; and what kinds of processing will be arrayed to convert the crude oil into the desired products. The many possible methods of processing include crude-oil distillation, hydrocracking, desulfurization, reforming, steam cracking, and catalytic cracking.

At every stage of designing a refinery, the process engineer balances competing needs. The cost of building the facilities must be weighed against the cost of operating them. Facilities that are less expensive to build are typically more expensive to operate. Gasoline of higher quality can be sold for a higher price than gasoline of lower quality, but the facilities required to produce gasoline of higher quality are more expensive to build and to operate. The cost of achieving ever-lower emissions to comply with environmental regulations must also be considered. Perhaps most critical is ensuring the safe operation of the refinery. Of course, human life is irreplaceable, and moral considerations are paramount. But we can also note that a design which kills people will also end up being hellishly expensive for the refinery owner.

The design of a low-cost, low-polluting, economically efficient, and consistently safe refinery is an intellectual exercise of the highest order—and great fun.

Engineering to the Rescue

For many decades, until the 1990s, many millions of tons of hydrocarbon emissions (some highly carcinogenic) were released into the air. That's a lot of bad emissions.

In 1994, I was doing consulting work for a company that owned a natural-gas processing plant. Natural gas, which is mostly methane, is extracted by exploration and production companies and delivered to customers by pipeline. The companies drill holes into rock that contains natural gas under pressure. The gas flows up the hole and through a pipe to the surface, from which it is piped to a natural-gas processing plant to be processed, compressed, and put in the pipeline for delivery to consumers.

When the gas is processed, contaminants like water, carbon dioxide, and hydrogen sulfide must be removed. If water isn't kept out of the pipeline, it can lead to corrosion and failures that result in massive towers of flame being spewed out of the ground, eruptions that sometimes kill people. The water contained in the natural gas itself will form natural gas hydrates when it freezes (at about 60 degrees Fahrenheit, not the 32 degrees at which pure water freezes). If allowed to form in a pipeline, solid natural-gas hydrates can also cause various problems—everything from plugging of the sensors used to control the pipeline to damage of pipeline compressors.

The most common device for removing water from natural gas is the glycol dehydrator, which uses triethylene glycol (TEG), a chemical with a great affinity for water, to do the trick. (Glycol dehydration was first used in the 1920s. The original agent of water removal, diethylene glycol, was eventually replaced by triethylene glycol.) When natural gas vapor interacts with the TEG liquid in a device called a contactor, the TEG absorbs water from the natural gas. The now-"dry" gas exits from the top of the contractor as the wet TEG liquid flows out from the bottom and into a reboiler. In the reboiler, the water captured by the TEG is boiled away, leaving the dry and intact TEG to be returned to the contactor. The recycled TEG now contacts more wet natural gas to remove more water. TEG is a very heavy liquid; it won't vaporize at all at the temperature at which water boils. So the same TEG continues to cycle endlessly through the system.

TEG also has a great affinity for benzene, toluene, ethylbenzene, and xylene (which the industry calls BTEX), of which natural gas contains trace amounts. Unfortunately, workers in the natural gas industry were for many years—about a century—unaware that natural gas contains these trace amounts of BTEX. TEG absorbs both BTEX and water from the gas. So when the reboilers of the glycol dehydrator boil away the water from wet TEG, in the absence of any further intervention they also boil away any BTEX that the gas contained.

For decades, then, the industry obliviously poured vast quantities of toxic BTEX into the atmosphere. During this period, the United States had about 30,000 glycol dehydrators processing the vast amounts of natural gas being produced in the country. The fact that benzene is a powerful carcinogen made such widespread emissions a public hazard.

Only in the 1990s did EPA officials and members of the industry become aware of the problem. In 1995, to help fix it, I started a small company to design and build glycol-dehydrator vent condensers. Such condensers receive the vapor stream of water, BTEX, and other substances being boiled away from the wet TEG in the reboiler; cool the vapors to within about 10 Fahrenheit degrees of the ambient air temperature; and collect the condensed liquids (mostly water and BTEX). The air flowing across the condenser tubes carries away the heat of condensation, and customers dispose of the condensed liquids that are collected.

The water and BTEX form separate liquid layers. The BTEX is sold and the water is usually sent away for treatment and disposal, since tiny amounts of BTEX do dissolve in the water and must be removed. In most cases, the revenue earned by selling the BTEX is enough to provide a good annual return on the investment in the condenser, sometimes a return greater than 100 percent.

The Case of the Deadly Dehydrator

In the mid 1990s, Company X discovered a large field of natural gas in central Texas. The gas was especially dry; i.e., it contained very little hydrocarbon that was heavier than propane and contained only small amounts of ethane and propane, which are light hydrocarbons a bit heavier than methane and always found in natural gas. Since the gas was so dry, the folks at Company X concluded that they would have no problem with BTEX. After all, BTEX is much heavier than propane. So there couldn't be any BTEX in their natural gas, could there?

To process the plentiful gas from their new field, the company installed a very large glycol dehydrator, but without also installing a vent condenser to control emissions from the reboiler. After all, only water was being boiled away. There couldn't be any BTEX being boiled away along with the water, could there?

Within a mile of Company X's brand-new, very large glycol dehydrator was a nice ranch owned by a member of the Texas legislature. As soon as the new glycol dehydrator began to operate, the rancher noticed a terrible smell. It was so bad that he, his family, and several ranch workers had to move out of the area.

I received a panicky call from Company X. They had shut down the field because they'd had to shut down their glycol dehydrator in obedience to the demands of all sorts of Texas government officials.

After performing exhaustive quantitative analysis on samples of the gas flowing from their wells, Company X found that although the gas contained very few heavy hydrocarbons, substantial parts per million of BTEX were indeed present in the gas. Given the very large quantity of natural gas being produced and treated, that amount of BTEX resulted in a great deal of BTEX being emitted by the reboiler.

The company asked me to deliver a rather large vent condenser as quickly as possible. Having obtained their gas analysis and the details of their glycol dehydrator design, we began work on the design immediately; and once we received the materials we needed, we worked around the clock to build the condenser. We gave Company X a fair price—one that took into account the costs and hassle involved in doing the job at breakneck speed.

The vent condenser was installed and the glycol dehydrator was restarted. The condenser functioned. The smelly BTEX emissions disappeared. Company X and the rancher were happy.

Counting Annual Emissions

The Environmental Protection Agency was all over the BTEX emissions problem—once they realized that this massive problem existed. The agency decreed that any glycol dehydrator emitting over 25 tons a year of hazardous air pollutants (i.e., BTEX) would be regarded as a major emitter of hazardous air pollutants. In addition to the direct cost of this designation in the form of fines, indirect costs would also be steep. Any company deemed a major emitter would have problems with investors and environmental activists. Nonmanagerial members of the company's board of directors, like college professors and investment professionals, could expect severe criticism from their colleagues.

Every company that operated glycol dehydrators had a new top priority: proving that the devices for controlling emissions installed on the reboilers of those dehydrators kept their annual hazardous air pollutant (HAP) emissions below 25 tons.

One way to control emissions is by using a flare. Simply burn all the hydrocarbons in the vent stream, turning them into water and carbon dioxide. There are two problems with that. A flare that destroys all HAP emissions is a pain to operate. And given the massive amounts of water in the vent stream, it is very costly to operate; a flare destroys the BTEX, which when preserved and collected can be sold for a lot of money.

Clearly, the ideal solution is a condenser that liquifies both the water and the BTEX of the vent stream (as well as other hydrocarbons in the mix). This is the best method, even taking into account the cost of treating and disposing of the contaminated water. But a condenser won't condense 100 percent of the BTEX, so how do we figure out exactly how efficient it is at doing so? We need to determine its efficiency across a year to make sure that annual emissions of BTEX do not exceed 25 tons.

For almost every glycol dehydrator, the most practical condenser is an air-cooled condenser. Most locations have no cooling water available, and installing a cooling tower to provide the water required to remove the heat from the condenser is usually impractical. This is why home air conditioning units use air-cooled rather than water-cooled condensers. Some large commercial air-conditioning units do use water-cooled condensers, but these are an exception.

When I started my company, nobody knew how to determine how efficient an air-cooled vent condenser is at recovering BTEX over the course of a year. But discovering a method of determining this was critical, since I wanted to sell air-cooled condensers to my gas-processing customers.

The efficiency of the device is affected by air temperature. If the outside air temperature is 125°F, your home air conditioner performs poorly; it cannot condense as much refrigerant with such hot air as it can with air at 90°F. The same is true of vent condensers for glycol dehydrators. Air-cooled vent condensers are more efficient on cool nights than on hot days, more efficient in winter than in summer.

How do you determine the annual efficiency of a condenser the efficiency of which changes hour by hour throughout the year? The first step—after asking the question—was to decide that it could be answered. Then I formulated the problem in my mind more or less as described above and told myself, "Okay, right brain, the problem is yours." A few days later, I awoke with the solution.

To define the climate where a vent condenser is located, obtain Weather Service data on air temperatures for a place nearby: the 8,760 hourly temperatures that have been recorded for each year you select. These records are available on DVD for locations all over the country and cover scores of years. (The Weather Service also keeps rainfall records, but rainfall amounts are not relevant to determining the efficiency of vent condensers.)

The temperature data enable you to create an annual-temperature histogram like this one, which I presented in a paper at a gas industry conference on toxic air pollutants.

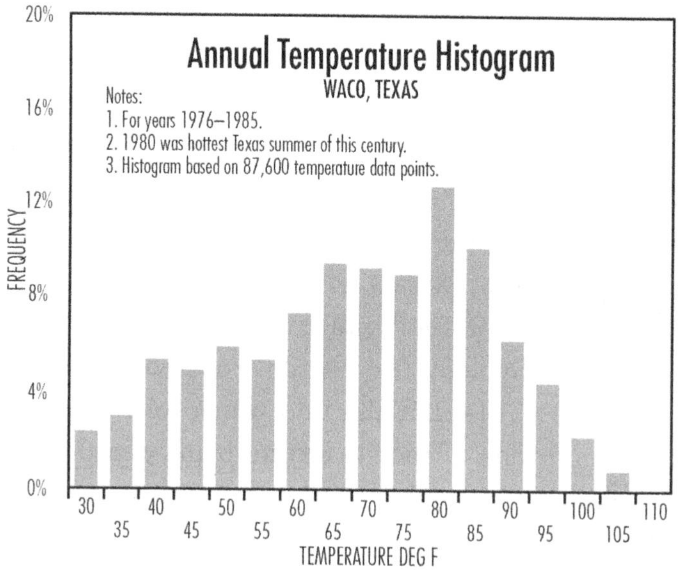

This histogram shows the part of the year during which the recorded temperature falls within each specified increment of five Fahrenheit degrees. For example, about 13 percent of time, the temperature is between 80°F and 85°F. Histograms based on increments of ten degrees or two degrees would look different. But five-degree increments are a reasonable basis for calculating annual HAP emissions; the amount of variation in natural gas analysis is such that a higher degree of precision is not useful.

A histogram of the air temperatures for a particular year at a given location looks much like the histogram of the air temperatures for any other year at that location. A location's climate is the location's climate. Year to year, climate hardly changes at all. You can confirm this by using Weather Service data to construct a whole bunch of annual temperature histograms for a few locations, comparing different years for each location.

Now that we have determined the annual air temperatures at the location of the condenser, in order to determine the temperature of the (mostly condensed) vent stream of the glycol reboiler, we must know how the condenser actually operates. The Gas Technology Institute developed a computer simulation (GRI-GLYCalc) into which we can input the composition of the gas stream being dehydrated and input data on the dehydration unit. This simulation determines the composition of the vent stream with high accuracy. Using the composition of the vent stream and the outlet temperature of the condenser, we can use familiar formulas to calculate how much BTEX remains uncondensed and is therefore emitted into the atmosphere at each outlet temperature.

With this information—which includes the quantity of heat released for each increment of cooling—we can design an air-cooled vent condenser and know what the condenser's outlet temperature will be for every air temperature. I designed my vent condensers to cool the outlet streams to within ten degrees of the highest expected air temperature. For example, if the maximum air temperature is 110°F, the maximum outlet temperature of the condenser is 120°F. At every lower air temperature, the outlet temperature will also be within ten degrees of the current air temperature.

This is because the greatest heat removal (the water condensation) occurs near the condenser inlet at the highest temperature inside the condenser. This ten-degree difference between the outlet temperature and the ambient air temperature is called the temperature approach.

To create a temperature histogram showing the outlet temperatures of the condenser, we shift everything in our air temperature histogram ten degrees to the right. This is not quite precise, since we do not want the condensed water to freeze if the weather is very cold. To keep the water from freezing, we control the outlet temperature, keeping it above 50°F. In cold weather, the dehydrator operator uses valves to temporarily disable parts of the condenser I designed. Emissions are vanishingly small when the temperature approaches 50°F.

In the latest version of GRI-GLYCalc, Version 4.0, the Gas Technology Institute includes many histograms of location temperatures in the United States and the means of choosing a condenser temperature approach and of calculating the annual emissions of your air-cooled vent condenser.

The developers of GRI-GLYCalc 4.0 have been kind enough to acknowledge me for suggesting "the methodology for estimating site-specific annual air-cooled condenser emissions based on ambient temperature frequency data." You can buy a copy of this marvelous software for $295 at the GTI page sales. gastechnology.org/000102.html.

Let's Frack

Methods of extraction designed by engineers continue to make our civilization possible.

Hydraulic fracturing or fracking is a method of extracting crude oil and natural gas by pumping water at very high pressure into rock deep formations to create crevices in the rocks that contain the oil and gas. The oil and gas flow into those crevices and into a well pipe to the surface, where it is processed and sold.

Socialists hate fracking because they hate oil companies and capitalism. They have also cast the oil companies as premier villains of "climate change." So the socialists hate anything that oil companies do.

Our civilization cannot exist without the production of oil and gas. No alternative sources of energy can do the jobs that oil and gas do. The anti-oil progressives are untethered to reality. They routinely complain about things without which civilization cannot function while offering no rational alternatives. ("Renewable energy" is not an alternative, only an empty slogan.) Fracking accounts for a large and growing proportion of the world's oil and gas production. Without it, the cost of these fuels would skyrocket.

Progressives contend that fracking will ruin groundwater. But much "groundwater" is "ruined" in its natural state. The groundwater where fracking takes place lies a mile or more underground. For over a hundred years, oil companies have recovered such groundwater along with the oil and gas, calling it produced water or brine. Produced water contains dissolved salts and gases, some of which are naturally radioactive. In other words, the water is already contaminated. It always has been.

This deep groundwater has been around for millions of years. It is separated by a mile or more of rock from the shallow aquifers that produce much of our drinking water. The same thick layer of rock also serves as a barrier between fracking-extracted water and the aquifers. So there is no "groundwater contamination." The claim is just a baseless talking point, at best a product of ignorance, at worst a lie.

Listening to socialists, you might suppose that the environment of any locale near extensive fracking operations must be greatly damaged. But consider the case of Fort Worth, Texas as one of many counterexamples. Fort Worth is a fine place to live. Yet it lies in the heart of Tarrant County, above the Barnett Shale, which contains large quantities of natural gas. Oil and gas companies have performed extensive hydraulic fracturing to recover this gas: there are about 15,000 fracked gas wells in the county, or about 16 fracked gas wells for each of the county's 902 square miles. All of this fracking has resulted in zero impairment of Fort Worth's quality of life.

There has been much anguish in England over the use of fracking. Lancashire County rests on shale that is rich in natural gas. The socialists sometimes say that even if Texas with its lower population density is suitable for fracking, England, being a more densely populated land, is not.

However, Tarrant County, Texas has a population density of about 900 persons per square kilometer; Lancashire has a population density of about 475 persons per square kilometer. Tarrant County with its 15,000 fracked shale gas wells has a population density almost twice as great as that of Lancashire.

It's okay to frack.

SEVEN
EDUCATION OF A WANDERING MAN

My desire to see new places and do new things shaped my life.

I was born in Houston, Texas. My father, who had dropped out of college during the Great Depression, was an electrician and then a supervisor in a chemical plant. My mother was an elementary school teacher. I spent my early years in Baytown, Texas, except for a year during World War II in the Los Angeles area, where my father was doing construction work on a new oil refinery.

After high school, I studied at Rice University in Houston, where I received a B.A. and a B.S. in chemical engineering. While I was attending Rice, my father was able to secure a summer job for me at the chemical plant where he worked. In addition to paying most of the costs for my education at Rice, including living expenses and lab fees (the school did not charge for tuition at that time), the work gave me a pretty good idea of how I might use what I was learning.

Growing up in Baytown, I had the chance to watch ships from all over the world, their home ports proudly displayed on their sterns, as they sailed up and down the Houston Ship Channel. Imagining the exotic home ports of all those ships made me want to go where they had come from. So before I graduated from Rice, I applied to the Officer Candidate School of the United States Navy. Unfortunately, I failed the required eye test, so I didn't make the cut.

About the time I was being rejected by the Navy, though, job offers were coming in from various companies. The most lucrative offer came from Exxon, for a job in their Texas refinery. But it was easy to turn Exxon down in favor of the international division of the Union Carbide Corporation (UCCI). I wanted to travel, and it looked like a job with Union Carbide would let me do that.

At first I was assigned to UCCI's engineering department in Houston. Not much travel yet—just the trip from my room at Rice University to a shared apartment in the Sin Alley area of West Houston. For about a year I designed heat-transfer equipment for UCCI facilities in Brazil, India, and Sweden. Then I was transferred to a team in Houston doing the process design for a chemical plant to be installed in Sicily. But the offices for the mechanical design phase of the project were in Milan, Italy; so a little less than two years after I had graduated from Rice, I was working in Milan and living in my own apartment at Viale Piceno 13.

Italy was wonderful. Milan was wonderful. The food was wonderful. I often walked from our office to the Galleria Vittorio Emanuele II, where I would eat lunch at Ristorante Biffi. I would sit at a table in the center of the Galleria, read the *International Herald Tribune*, watch the world go by. Milan was very central. My apartment was about a 45-minute drive from the Swiss border and about a three-hour drive to Venice or to Florence. Every weekend was a chance to see more of Italy. During my two-week vacation, I made sure to drive to Berlin and visit both sides of the Berlin Wall.

Unfortunately, my stay in Italy lasted only a year or so; technological improvements rendered obsolete the plant design we were working on. So it was back to Houston after an all-too-short year in Italy, including only a couple of business trips to Sicily. The highlight of one of the trips had been a meeting with local dignitaries at their club in a medieval castle on Ortygia Island, a part of Syracuse. A corner of the club bar featured a hole in the floor through which, centuries earlier, unfortunates had been pushed to plummet to the dungeon below.

Back in Houston a new project awaited, a chemical plant to be built in Puerto Rico. I had been working on the process design of the new plant for about a year when I met Gayle on a Galveston beach. Within a year we were married, and two months later we took a delayed honeymoon in Jamaica, Martinique, Guadeloupe, Antigua, and St. Thomas on the way to our new home in Ponce, Puerto Rico.

Our year and a half there was filled with hard work and hard play. In addition to learning quite a bit about the culture—Gayle became fluent in Spanish thanks to a Berlitz course and lots of practice—we also learned about each other as we traveled all over the island, rode the ferry to St. Thomas, and flew to other islands. Union Carbide gave expatriates like us a trip back to the States once a year, but we traded in our tickets and flew to Spain instead for our two-week holiday. We loved Franco's Spain. The people were friendly and helpful, the food was great, and the prices were low. A room at lovely hotel in Murcia cost us $2 a night—although we did have to go for a short walk down the hall to get to the bathroom. The glorious and nearly deserted Alhambra palace and fortress in Granada was the highlight of the trip.

Our time in Puerto Rico was incredibly busy. I worked long hours to ensure that the plant design prepared in Houston was faithfully executed, which included preparing the formal change orders necessary to correct occasional design errors—usually, errors made by the engineering contractor as he converted the process design into drawings of the equipment foundations, piping elevations and plans, and so forth.

Once our Puerto Rico sojourn was over, it was back to Houston, where things got really interesting. Within five years we had bought a house and had our two children.

Also during this time, the engineering department of Union Carbide's international division ceased to exist, having been subsumed into the engineering department of the company's chemicals and plastics division. It was a sad event for us UCCI people. But life goes on. The shuffling meant that I had a new boss and that instead of working only on new plants outside of the United States, I would now be dealing with the company's American business. I was suddenly working in a new and much more staid culture, albeit one with many more technical resources. At UCCI, we had always regarded the domestic engineering department as somewhat stuffy; they, in turn, had regarded the engineers at UCCI as rather weird.

Be that as it may, I was soon assigned to the engineering group for the low-density polyethylene (LDPE) business team. This fascinating work involved business planning as well as chemical engineering design. For some years, Union Carbide had been the largest manufacturer of LDPE in the world. One of the less happy consequences of this dominance was the excessive cautiousness of management, the attitude of "Don't change things! This is the way we have always done it!" Thanks to years of such timidity, the company's technology for producing LDPE had become outdated, with the superior technology of our competitors giving them a competitive edge. Union Carbide's corporate officers were not oblivious of the situation. Despite their cautiousness, they knew that the company's LDPE business needed to be more innovative and even daring. So the new marching orders were: "Come up with new LDPE technology that will help us beat the competition!"

One of my tasks was to help assess the technology of our competitors. Although we thought that we had a pretty good idea about what they were doing, we also knew that we were guessing to an extent. One day my boss stressed how urgently his boss's boss wished that we could learn more about the competition. What we should do, I suggested, is rent a helicopter to fly over the new Exxon polyethylene plant in Baton Rouge and take pictures. I thought I was just musing, engaging in a theoretical exercise, until I was given precisely this task. I chose a 35 mm camera loaded with fast film and a 300 mm lens for that exciting helicopter ride. The pictures we developed were magnificent, and they pretty much confirmed what we had suspected about Exxon's technology.

Our escapade led to a minor intercorporate incident. The Exxon people on the ground managed to record the tail number of our helicopter, and someone traced it back to Union Carbide. The CEO of Exxon then called the CEO of Union Carbide. I was not told the gist of that conversation, but my immediate superiors were very nervous for a while. So far as I knew, though, there were no further consequences.

Union Carbide's chemicals and plastics division suffered from a continuing irritant: the existence of an engineering office in Houston. Historically, the division's R&D and engineering design had been centered in Charleston, West Virginia. For whatever reasons, the managers of the engineering division decided that the Houston engineering office must be shut down and its personnel transferred to "the tech center" in Charleston. Not long before, a major building project had greatly expanded the office space there, so that we could all be nicely accommodated.

Virtually everyone in the Houston office moved to West Virginia. Everything worked out well for Gayle and me. The work went on as before, and we found a house high on a hill looking down on the Kanawha River Valley. It was only a short walk through the woods and down the hill to a nearby elementary school. And the Charleston area was a cultural oasis of West Virginia. Gayle took a job as church choir director and organist, was a volunteer music teacher at the nearby Weberwood Elementary School (which the kids attended), and managed to commute to Marshall University in Huntington, WV to obtain a masters degree in organ performance.

Life was good. My own work became more and more interesting. The departments of polyethylene engineering (where I worked) and of polyethylene research and development had begun an exhaustive study to discover the ultimate cutting-edge technology for manufacturing low-density polyethylene. No idea, however unlikely, was ruled out. As it turned out, the ultimate technology was soon within our grasp, for Union Carbide had been developing a marvelous gas-phase technology for manufacturing *high*-density polyethylene. If we could adapt that technology to the production of *low*-density polyethylene, we would own the world-beating technology.

We did. We developed it. The effort was all-out. It seemed very likely that this technology would sweep aside most of the competition and eventually account for most LDPE production worldwide. There was a problem, however. The investment required to install the facilities that could capture all of that business was more than management wanted to spend. They determined that the company could make just about as much profit if, instead of building that capacity, we licensed our LDPE production technology to others. We'd let them spend the money to build the new capacity as we collected generous licensing fees.

Once Union Carbide decided to license the UNIPOL technology, a team was assembled to manage this new endeavor. The LDPE business-licensing team included people from sales and marketing, production, R&D, and engineering. I was promoted to engineering manager for polyethylene licensing. We would license UNIPOL for the production of both low-density polyethylene and high-density polyethylene.

Our job was to market and sell UNIPOL technology, and, once it had been sold, to transfer that technology to the business customer in a way that best served the interests of both Union Carbide and the customer. I was primarily responsible for the transference, since the customer would be implementing the technology by designing and building a new LDPE facility. I was provided with subordinates who would manage the transfer to each particular customer. Perhaps it goes without saying that once a company developed a serious interest in licensing UNIPOL, to obtain more detailed information it had to sign a confidentiality agreement guaranteeing that the information would not be disclosed to any third party nor used in their own businesses except as permitted by the terms of our licensing agreement.

All of this seemed straightforward. But it soon became clear that I would also be performing another task. With backgrounds mostly in R&D, none of the salesmen in our marketing and sales group had any experience with the designing and building of plants. So they were unable to answer the many questions that our customers were asking about this subject—including, for example, detailed questions about the capital costs of facilities. Well, there was an obvious remedy. Me. I began to regularly join the salesmen's later meetings with customers so that I could share my engineering expertise. Then I was asked to join their early sales meetings as well.

One advantage of UNIPOL was that UNIPOL facilities were much safer to operate than the facilities for competing processes. I had substantial experience with process hazard analysis, and the safety benefits became a feature of our marketing.

Then a third task was dumped in my lap. Especially in the case of our less technically adept customers, it was desirable to be able to offer the customer a complete package: i.e., to provide him with not only the technology but also a detailed mechanical design for the new UNIPOL facility *and* construction of that facility. Then the customer wouldn't have to worry about building it. We would build it for him with the help of an engineering firm that Union Carbide would take on as a contractor.

That contractor would have the expertise required to complete the mechanical design, acquire all necessary materials, and manage the construction work. A British engineering and construction firm called Constructors John Brown (CJB) filled the bill, and my new task was to serve as the primary liaison with this firm. Its main office was in Portsmouth, England, so I became well acquainted with the Portsmouth area. I also came to know the CJB personnel quite well, and my family eventually exchanged visits with the family of one of their managers.

We had picked CJB in part because of their success in selling their design and construction services to the Soviet Union, and one of our early joint sales efforts targeted the Soviet Union. In 1976, a UNIPOL sales rep and I traveled to Moscow with a CJB sales team. This was the trip during which I witnessed the windshield wiper caper that I describe in chapter three, with the driver carefully removing and hiding the windshield wipers after parking his car so that thieves could not so easily make off with an item in socialism-induced short supply.

In the Soviet Union, I observed socialism in its most extreme form. The dreariness was not entirely unleavened. Our most entertaining meeting was held at Techmashimport, the Soviet organization responsible for acquiring foreign (i.e., Western) technology. Its managing director was Comrade Petrov, certainly a KGB officer, and likely at least a colonel. His office entrance had two oppositely opening and padded doors, which when closed left a space of only an inch or two between them. Inside the comrade's office and above the inner door was a red light, and any time either of those two entrance doors was open, the red light above the door would blink on. Sitting in that office I felt a little like James Bond.

Because Union Carbide did not want to license UNIPOL to our domestic competitors, all of my sales trips were to places outside of the United States, usually to Western European countries or Japan. Our most fascinating trip was to communist China. We visited in 1978, not too long after the Cultural Revolution that Mao Zedong had inflicted upon the Chinese people. We saw indications of how totally many of the Chinese people had come to reject the basic ideas of Communism, even though they were still under the power of a government that, if no longer full-on communist, was still in many respects totalitarian.

In dealing with our Chinese hosts, I was startled to observe how similar the Chinese people were to Americans. After never really connecting with the Japanese after several trips to Japan, I found that the Chinese seemed to look at the world much as we did, and it was a wonderful experience. I had the chance to give presentations in English using slides and graphs in Mandarin. Every few seconds I paused to permit time for my words to be translated. Despite the translation routine, the question-and-answer sessions worked quite well.

The late 70s was an unusual time for China. The country was beginning to open up to the world for the first time in decades, and it was not really prepared for the sudden surge in visitors. It meant, for one thing, a scarcity of nice hotel rooms in Beijing, obliging us to leave Beijing and conduct many of our meetings in Tsing Tao, in the province of Shandong on the Yellow Sea. The coastal area there reminded me of California.

Tsing Tao had been a German enclave in China and had many legacies from the German days. We lodged in a guest house that looked as if it had been transplanted from turn-of-the-century Frankfurt. Three very nice meals were served daily by a staff of nine who took care of our group of six. We were the only guests. The meals were paid for on a per diem basis—except for the beer, which was extra. Our hosts assured us that the excellent Tsing Tao beer was still made according to the traditional German beer-making rules known as *Reinheitsgebot*. During our week-long stay, we paid for 78 liters of beer.

When we walked around Tsing Tao, the local Chinese seemed even more interested in us than we were in them. Wherever we went, we attracted a crowd of curious onlookers, very often standing in a circle around us twenty or thirty deep. We were told that some of the younger people there had never seen a Westerner before. It was a quite an experience.

The job was more than satisfying my desire to travel. Gayle kept track and found that in one year, I was out of town roughly two thirds of the time. That was a bit much. I loved my job, but I had been doing it for over five years. A change seemed appropriate. And although we had enjoyed living in West Virginia, it was a bit insular, and we thought that the children might benefit from living in a different place.

I began looking at other possibilities around the company. We visited the New York office of Union Carbide and considered buying a home in Connecticut. I also began looking for a job outside of the company. Eventually, after a lot of phone calls, I accepted a job with Texas Eastern Corporation as the project manager for a massive new facility for manufacturing synthetic fuel. Fluor Corporation had been selected as the job's engineering contractor, so I moved with the family to Irvine, California, then home of Fluor's main office.

For the children, moving from West Virginia was something of a culture shock. For Gayle and me, it was another interesting interlude. The job in Irvine lasted only a year. The expected continuous run-up of crude oil prices did not transpire, so that the economic basis for producing synfuels ceased to exist. The good news for Texas Eastern was that many of the costs of the project had been borne by the federal government.

In any case, it was back to Houston for us after only a year in lotus land. I became a manager in Texas Eastern's corporate engineering division and later became involved in the company's real estate division. There is nothing especially notable to report about my time there other than my discovery that the real estate business was not my cup of tea. Even though the real-estate operation contributed little to the company's bottom line, it was a pet project of upper management; it gave company officers a lot of face time in the local community.

In 1989, Panhandle Eastern acquired Texas Eastern. Major staff cuts were imposed, and I was one of many offered an early retirement package. The offer was extremely generous. For example, it included continuing free health insurance for retirees and spouses as well as the opportunity to choose a lump sum payment of retirement benefits in lieu of an annuity. There was no way I could reasonably turn it down, so suddenly I was in early retirement and, not willing to remain retired, looking for work.

A good five of the recent years in Houston had been spent largely marking time at work. But Houston itself had provided an abundance of engaging opportunities. For example, I earned an MBA during those years. I also took a course in Writing the Novel taught by Venkatesh Kulkarni at Rice University, a wonderful experience that lasted several years. Kulkarni, author of the novel *Naked in Deccan*, was a marvelous teacher. And I learned how hard it can be to try to write a good novel.

After some months of job hunting, I was fortunate to find work as a project manager at Howe-Baker Engineers in Tyler, Texas. Tyler, in the northeastern part of the state, was not far from the home of my father-in-law, who was elderly and would benefit from our proximity. I worked at the firm for only a couple of years before getting laid off. Maybe I wasn't a good fit. But once again, luck was with me. A neighbor in Tyler ran a different, even smaller engineering firm, Gardner Engineering, which provided electrical engineering services. Several of his clients were in the gas processing industry, and he thought that he might expand his range of services to include chemical engineering consulting.

Some years before, I had obtained a license as a professional engineer. Although I hadn't really needed it for my work at Texas Eastern, I'd figured "Why not?" (Perhaps suspecting that it might one day come in handy.) With the license in hand, pitching my services as an engineer was straightforward and in compliance with professional engineer licensing laws.

Fortunately, Burlington Resources needed my services at once. Their large central Texas gas processing plant had possibly run afoul of some new federal government safety regulations, and it was my job to sort out that plant's technical documentation in order to prove that the plant was complying with the new regulations. I had to make sure that the client's actual, installed plant facilities matched the documentation—a complicated task that would take quite a while.

Some months later, as I was finishing this first job, a gas processing plant owned by Occidental Petroleum Corporation in Far West Texas began having a problem with their operations. Since we had a consulting contract already in place with this company, it was amended to include provision of my services as a chemical engineer.

While analyzing the processing problem, I became intimately familiar with the plant's glycol dehydration system. Solving the problem required modifying the operation of their dehydrator and controlling the resulting greatly increased emissions from the dehydration system's reboiler. Of necessity, I became immersed in the whole subject of glycol dehydrator emissions control. I also designed a new system to control the emissions of toxic air pollutants from that glycol reboiler.

At the time, glycol dehydrator emission control was a red-hot topic. The Gas Research Institute (today the Gas Technology Institute) had established a series of annual conferences on gas industry air toxics, and I presented papers at two of these conferences. The first included the method for calculating annual air toxics emission control described in chapter six. At this conference, I met many people who 1) wanted to understand this emissions control technology and 2) needed to acquire the equipment that would do the controlling.

Here was a clear opportunity. All I had to do was provide vent condensers for the gas processing industry.

I approached my boss, the owner of Gardner Engineering, and suggested that we go into the business. He demurred, but asked, "Why don't you do it yourself?" The suggestion surprised me, but I decided to explore the possibility. I looked for a welding shop that could convert my designs into saleable hardware.

Thus was born Dehy Condensers, Inc., a business that lasted about eight years. Dehy was owned by the owner of the welding shop, and we concluded an agreement for my services. In return for about a fifth of the sales revenue, I provided the sales and marketing for the operation, and I designed the equipment. I was the research and development, engineering, and sales and marketing departments. We turned a tidy profit. It was great fun, but it was also a hell of a lot of work.

I had become a full-blown entrepreneur. I created

a product for which customers were pleased to pay substantial sums in order to solve a pressing problem. I also supplied the customer with a document attesting to the degree of annual emissions control that my product provided for the customer's glycol dehydration system. The document bore my stamp as a professional engineer, which gave it credibility with state and national environmental protection agencies.

We had a good run, but eventually I grew tired of the workload involved in running the business. And by then the grandchildren had begun arriving. So we decided to make another change, and I finally and permanently retired. I shut down the business and we moved to Nashville, Tennessee, where the grandchildren and the action were. We later made one more move, to our current home in Brentwood, just south of Nashville. It is a very pleasant place to live, and although we occasionally miss Texas, we are content.

EPILOGUE

The president for life of Russia, Vladimir Putin, once said that the demise of the Soviet Union was the "greatest geopolitical catastrophe" of the twentieth century. In his view, the seventy years of the existence of the Soviet Union represented the complete and glorious flowering of socialism.

Putin's lament is a ringing endorsement both of the socialism that was and the socialism that will be. His words reveal the enduring power that the ideas of socialism has over men's minds. The fact that it is nonsense—very destructive nonsense—in no way diminishes this power, which derives from socialism's appeal to all manner of human weakness and irrationality.

If you envy people who have more than you, socialism has something for you.

If you cannot stand the imperfections in the world around you, socialism has something for you.

If you are obsessed with race, gender, the homeless, climate change, the environment, or some other fashionable cause, socialism has something for you.

If you long for a king or führer to tell you what to do, or if you imagine yourself as part of a ruling aristocracy telling others what to do, socialism has something for you.

If you are a religious fanatic, socialism has something for you.

If you believe that the end of everything is approaching, socialism has something for you.

If you have the mental illness known as Munchausen syndrome by proxy, socialism has something for you.

Capitalism and freedom do not lend themselves to May Day parades, mass demonstrations, and blinding emotions. The fact that capitalism is good enough to achieve all the great things that it does achieve, albeit imperfectly, does not exert the same emotional appeal over gullible minds that an imaginary socialist perfection exerts.

Certain gullible people will always find the lies of socialism more persuasive and attractive than the unquestionable success of capitalism.

The only thing that capitalism, true democracy, and individual freedom have to recommend them is the fact that they work brilliantly. They keep us alive and enable us to flourish.

That's it. But that's enough.

ACKNOWLEDGEMENTS

I must first of all thank my wife, Gayle, the hardest-working person I have ever known. Gayle has been a church organist and choir director, a real estate agent, book club leader, master gardener, painter, quilter, knitter, and seamstress. There seems to be nothing she cannot do when she puts her mind to it. Her example is inspiring. Without Gayle, my life would have been less happy, more dull, incomplete.

I also owe a debt of gratitude to Joost Hudig, Jan Lowey-Ball, and Professor Riki Kobayashi, mentors and good examples.

I appreciate my mother's love of words, one of her gifts to me.

I am proud of our children, Wendy and Mark. My hope is that each is becoming a better person than I am. And thanks to Mark for helping me with the book's graphics.

Finally, thanks to my editor, David M. Brown. I do now appreciate the value of a skilled editor.

INDEX

A

Age of Reason 37
Alexander, Professor Larry 26
Althusser, Louis 50, 73
American Foundation for Suicide Prevention 19
An Act Declaring the Rights and Liberties of the Subject and Settling the Succession of the Crown 36
Analysis of UK Wind Power Generation: November 2008 to December 2010 114
Apocalypse of Peter 107
apostle Paul, the 106
Aran Isles 143
authority/subversion foundation 13
autobahn 47
Azolla 109, 110

B

Beijing 7, 172
Bell, Derrick 88
Belt and Road Action Initiative 47
Bergoglio, Jorge Mario 69, 70
Berlin 7, 163
Black Codes 77
black privilege 87
Bolivarian Revolution 21, 41, 50, 55
Book of Revelation 107
Brezhnev 7
Brown, Michael 90
Bruno, Giordano 82
BTEX 148, 149, 150, 151, 152, 153, 156
Buffet, Warren 64

C

Campus rape culture 22
care/harm foundation 13
Castro 40, 55, 62
Chavez, Hugo 20, 21, 40, 41, 62
Chinese Communist Party 46, 49, 62, 69, 96
Chinese Exclusion Act 88
Christianity 96, 98, 100, 131
Chronicle of Higher Education 23
climate change 105, 107, 108, 110, 119, 158, 179
clockwork universe 54

Comparing the Costs of Intermittent and Dispatchable Electricity Generating Technologies 117
Counter-Enlightenment 22, 25, 50, 82, 102
COVID-19 11
critical race theory 87, 88

D

Damore, James 83
da Vinci, Leonardo 123
Debs, Eugene 61
Declaration of Independence 33, 53
de Lévis, Pierre-Marc-Gaston 82
Deng Xiaoping 45
denier 52, 82, 119, 120
dispatchable power 117
divine right of kings 49, 61, 63
Domesday Book 32
Down's syndrome 79
Drawing on the Right Side of the Brain 127
Duke lacrosse 94
Duke University 94, 95
Duranty, Walter 42

E

Eco, Umberto 52
Edwards, Betty 4, 127
el-Tayeb, Sheikh Ahmed 70
Enabling Act of March 1933 70
Engels 50, 54, 86
English Bill of Rights 37
Enlightenment 22, 25, 33, 37, 50, 51, 82, 83, 97, 98, 102
Entartete Kunst 125
Environmental Protection Agency 151
Evergreen College 27
Exxon 118, 119, 162, 166

F

Factitious Disorder Imposed on Another (FDIA) 12
fairness/cheating foundation 13
false consciousness 73, 74
feedback control 57, 59, 60
feed-forward control 57, 58
female genital mutilation 18
Ferguson riots 90
feudalism 49, 51, 57, 61, 63, 71, 72
Feynman, Richard 4, 44
fiefs 61
five-year plans 55, 56, 58, 124
Fordham University's Internet Medieval Source Book 71
fracking 158, 159, 160
French Revolution 50

G

Galilei, Galileo 82
Gas Research Institute 177
Gas Technology Institute 156, 157, 177
Gates, Bill 64
general relativity 13, 54
German Centre Party 70
Ghost Ranch 128, 131, 132, 136, 137, 140, 141

Gibson's Bakery 92, 93
glycol dehydrator 148, 150, 151, 152, 153, 176, 177
Goebbels, Joseph 65
Google 83
Great Depression 46, 161
Great Leap Forward 7, 48
GRI-GLYCalc 156, 157
Guardian, The 18

H

Haidt, Jonathan 4, 12
Hands up! Don't shoot! 91
Harvard Crimson, The 84
Harvard University 29, 79, 80, 84, 86, 88
hate speech 26, 75, 102, 103
Hawthorne effect 138
Hayek, F. A. 4
hazard analysis 8, 170
Heterodox Academy 28
histogram 154, 155, 157
Hitler 40, 46, 48, 55, 64, 65, 66, 67, 68, 70, 86, 124, 125
Hoffer, Eric 5, 67, 120
Hollander, Paul 5
homeless 10

I

interglacial 111, 112
International Herald Tribune 163
Islam 71, 96, 97, 98, 99, 100, 101, 102

J

James, Clarence 93
Jesus of Nazareth 69, 106
John Muir Trust, The 114
Joskow, Paul L. 117
Jyllands-Posten 99

K

Kampf, Louis 74
Katyn Forest 67
Kennedy, Anthony 103
Kennedy, John 18
Keystone Pipeline 120
KGB 7, 56, 171
Kipnis, Professor Laura 22
Konzentrationslager 68
Kopel, David 77
Koran 96, 97, 99
Ku Klux Klan 77, 78
Kulkarni, Venkatesh 175

L

L'Amour, Louis 4
lebensraum 47
Left-brain 126, 129
Lenin 61
levelized cost of energy 116, 117
Lincoln, Abraham 34, 75
lobotomy 18
López, Ian Haney 88
loyalty/betrayal foundation 13
Ludlow, professor Peter 23
Lysenkoism 17, 18
Lysenko, Trofim 17

M

Magna Carta of 1215 36
Malaysian Airline Flight 370 19
Mangum, Crystal 94, 95
Mao Zedong 7, 46, 61, 69, 171
Marx, Karl 2, 50, 52, 54, 61, 67, 68, 86, 97, 106
Max Planck 130, 131
Mein Kampf 65
Me Too movement 24
Michelangelo 123
Middlebury College 26
moral foundations 12, 13, 14, 15
Moral Foundations Theory 13
Moscow 7, 42, 55, 56, 124, 171
Munchausen syndrome by proxy (MSBP) 12, 16, 17, 18, 20, 21, 24, 25, 180
Murray, Charles A. 25, 84
Muslims 45, 96, 97, 99, 100, 101, 102

N

Naked in Deccan 175
National Basketball Association 79, 87
National Football League 87
National Gay and Lesbian Task Force and National Center for Transgender Equality 19
National Socialist German Workers Party 64
Neue Rheinische Zeitung 68, 86
New Testament 98, 99, 107
Newton, Isaac 54
New York Times, The 42
Niemöller, Martin 41
Nifong, Mike 95
nine-dash line 47
nomenklatura 62

O

Obama, Barak 101
Oberlin College 92, 94
O'Keefe, Georgia 131
One Day in the Life of Ivan Denisovich 125

P

Pacelli, Eugenio 70
Pachai, Sundar 83
Page Act 87
Petition of Right 36
Philadelphia Inquirer 26
pilot suicide 19
placebo effect 138, 139
Pope Francis 69, 70, 96
Pope John Paul II 69, 70
Pope Pius XII 70
princelings 62
Prophet Muhammad 99
Pulitzer Prize 42

Q

quantum mechanics 13, 54, 130

R

Raimondo, Meredith 92
Reformation 98, 100, 101
Reichskonkordat 70
Reinheitsgebot 173
Renaissance 97, 98, 123
Rice University 161, 162, 175
right-brain 126, 127, 129, 130
Rolling Stone 93
Rozado, David 28
Russians Hungry, But Not Starving 42

S

safe spaces 22
sanctity/degradation foundation 14
Sanders, Bernie 75
San Francisco 10
SAT 85, 86
Savonarola 83
scientific method 13, 80, 81, 119
Scruton, Roger 4
Second Amendment 76
Shah, Zaharie Ahmad 19
Shakespearean sonnets 129
Sharia law 98, 99, 101
Smith, Adam 32, 45
Solzhenitsyn, Aleksandr Isayevich 125
Southern Poverty Law Center (SPLC) 26
Soviet Gulag 42
Soviet Union 7, 17, 40, 52, 55, 58, 62, 63, 64, 68, 69, 75, 120, 125, 170, 171, 179
Stalin 40, 41, 52, 53, 55, 64, 67, 68, 120, 124
Stanger, Professor Allison 26
Stewart, Potter 126
Strasser, Gregor 65
Stuart Young Consulting 114
Sunday Times of London, The 17

T

Taney, Supreme Court Chief Justice Roger 77
Tan'gun 63
Techmashimport 55, 171
The Bell Curve: Intelligence and Class Structure in American Life 25
the big lie 119, 120
The First Circle 125
The Gulag Archipelago 125
The Righteous Mind 4, 12
thermostats 60
The True Believer 5, 66, 120
The Truth About Gun Control 77
The Wealth of Nations 32
Third Reich 40, 55
Transgender 1, 16, 17, 18, 19, 20, 75
trigger warnings 22
Tsing Tao 7, 172, 173

U

Uighur Muslims 45
Ukraine famine 42
UNIPOL 55, 168, 169, 170, 171
University of Anglia 108

Unwanted Advances: Sexual Paranoia Comes to Campus 24
Ur-Fascism 52
U.S. Constitution 34, 35, 36, 89, 101

V

vassals 57, 61
Vermeer 141
Viner, David 108
Voltaire 39

W

Wax, Professor Amy 26
Weinstein, Professor Bret 27
white privilege 85, 86, 87
William the Conqueror 32
Wilson, Darren 90
windshield wipers 56, 59, 171

X

Xi Jinping 49, 69

Y

yellow peril 87, 88
yellow privilege 86, 88

Z

Zuckerberg, Mark 64

Made in the USA
Monee, IL
13 September 2026

62981966R00108